THE GREEN HOME: A DECISION MAKING GUIDE FOR OWNERS AND BUILDERS

LYNN UNDERWOOD

Australia • Brazil • Japan • Korea • Mexico • Singapore • Spain • United Kingdom • United States

The Green Home
A Decision Making Guide for Owners and Builders
Lynn Underwood
Vice President, Technology Professional Business Unit: Gregory L. Clayton mai

Acquisitions Editor:
Ed Francis
Development:
Nobina Chakraborti
Marketing Director:
Beth A. Lutz
Marketing Manager:
Marissa Maiella
Production Director:
Carolyn Miller
Production Manager:
Andrew Crouth
Content Project Manager:
Brooke Greenhouse
Art Director:
Benjamin Gleeksman

Library of Congress Control Number: 2009935483

ISBN-13: 978-1-4283-7709-7
ISBN-10: 1-4283-7709-3

Delmar
5 Maxwell Drive
Clifton Park, NY 12065-2919
USA

Cengage Learning is a leading provider of customized learning solutions with office locations around the globe, including Singapore, the United Kingdom, Australia, Mexico, Brazil, and Japan. Locate your local office at:
international.cengage.com/region

Cengage Learning products are represented in Canada by Nelson Education, Ltd.

Visit us at **www.InformationDestination.com** For more learning solutions, visit **www.cengage.com**

Purchase any of our products at your local college store or at our preferred online store **www.ichapters.com**

NOTICE TO THE READER

Cengage Learning and ICC do not warrant or guarantee any of the products described herein or perform any independent analysis in connection with any of the product information contained herein. Cengage Learning and ICC do not assume, and expressly disclaim, any obligation to obtain and include information other than that provided to it by the manufacturer. The reader is expressly warned to consider and adopt all safety precautions that might be indicated by the activities described herein and to avoid all potential hazards. By following the instructions contained herein, the reader willingly assumes all risks in connection with such instructions. Cengage Learning and ICC make no representations or warranties of any kind, including but not limited to, the warranties of fitness for particular purpose or merchantability, nor are any such representations implied with respect to the material set forth herein, and Cengage Learning and ICC take no responsibility with respect to such material. Cengage Learning and ICC shall not be liable for any special, consequential, or exemplary damages resulting, in whole or part, from the readers' use of, or reliance upon, this material.

Printed in the United States of America
1 2 3 4 5 6 7 12 11 10 09

BRIEF CONTENTS

CONTENTS

FOREWORD

Not having a guide to understand and quantify the issues of sustainability, what constitutes a "green building", and how to achieve energy efficiency in our buildings has perplexed the building community for some time. For homeowners this has been a complex and confusing issue.

The Green Home: A Decision Making Guide for Owners and Builders responds to these questions and takes a fresh look at the issues of what makes a green building, how we can achieve energy efficiency in buildings and what it means to have sustainable construction. While the author is a noted code official, his qualifications in the field of construction provide a unique and unbiased perspective in understanding the dynamics of these issues and how they impact the built environment.

Whether you are a new homeowner or are seasoned in ownership, this book provides a knowledge base to equip the reader with the concepts of green building and the premise of constructing a sustainable home or building. *The Green Home* is a guide for the individual, and our society as a whole, on construction, operation and maintenance. It is the complete package. Congratulations, Lynn on this accomplishment. Your work provides vital information in a time when broader understanding is needed and required on this most expansive concept.

Henry L. Green

ABOUT THE INTERNATIONAL CODE COUNCIL

The International Code Council (ICC) is a nonprofit membership association dedicated to protecting the health, safety, and welfare of people by creating better buildings and safer communities. The mission of ICC is to provide the highest quality codes, standards, products and services for all concerned with the safety and performance of the built environment. ICC is the publisher of the family of the International Codes® (I-Codes®), a single set of comprehensive and coordinated model codes. This unified approach to building codes enhances safety, efficiency and affordability in the construction of buildings. The Code Council is also dedicated to innovation, sustainability and energy efficiency. Code Council subsidiary, ICC Evaluation Service, issues Evaluation Reports for innovative products and Reports of Sustainable Attributes Verification and Evaluation (SAVE).

Headquarters: 500 New Jersey Avenue, NW, 6th Floor, Washington, DC 20001-2070
District Offices: Birmingham, AL; Chicago. IL; Los Angeles, CA
1-888-422-7233
www.iccsafe.org

This book was written as a guide for building a green home. As such, it refers to the 2006 edition of the International Residential Code (IRC). While the 2009 edition of the IRC has just been published and is available, it was not complete during the development of this book. There are some differences in the newer edition, especially for energy efficiency. Because of this, you are encouraged to purchase the 2009 IRC, especially if it has been adopted in your jurisdiction.

Chapter 1

GREEN HOMEBUILDING AND SUSTAINABLE DESIGN

THE ENEMY

From an environmental perspective, we have met the enemy: us. We are responsible for our environmental predicament.

Everyone wants and needs a home to live in—something that is safe, is structurally sound, has a sanitary water supply and refuse system, and includes some expected conveniences. We want a home that fits our needs, is pleasing to the senses, and is in a decent location. A home should be an investment to pass along as an inheritance to our children, a home that is solid and that will last. The problem is that when such a house is built, a corresponding inheritance is passed along that is not so readily apparent. Building such a house can have a negative impact on the environment created through construction and the demand for energy over its lifetime.

This problem covers many areas: wasteful and irresponsible consumption of energy and materials during construction; harmful effects on our planetary environment; depleted, nonrenewable natural resources; and adverse health effects on its owners. These are all caused by the choices made during the design of a new home and the actions taken during its construction. In addition, the use (or misuse) of fossil-fuel energy leaves a footprint on our environment. Carbon dioxide is released when hydrocarbon fuel is burned to provide several forms of energy. This carbon dioxide rises, entering the upper atmosphere, and rests in a blanket around the Earth. Each resident in the 100 largest U.S. cities is responsible for about 5,000 pounds of carbon dioxide. Since the advent of fossil fuels, coal and oil burning has caused an increase of these heat-trapping greenhouse gases in the atmosphere. In modest amounts, gases such as these in the upper atmosphere are necessary to life on Earth. However, in higher concentrations they retard heat from escaping to space, much like glazing in a greenhouse. These gases allow the sun's radiant heat to enter the atmosphere and prevent convective heat from leaving it. Although accurate recordkeeping for temperatures is recent (around 1850), during this time, our planet's temperature has climbed. According to data from the National Oceanic and Atmospheric

Administration (NOAA) and National Aeronautics and Space Administration (NASA), the Earth's average surface temperature has increased about 1.2°F in the last 100 years. The warmest 8 years on record have all occurred since 1998, with the warmest year being 2005.[1] Most of the warming temperatures in recent decades very likely have been the result of human activities, including energy usage. The Environmental Protection Agency's (EPA) position on global warming includes caution. "If greenhouse gases continue to increase, climate models predict that the average temperature at the Earth's surface could increase from 3.2 to 7.2°F above 1990 levels by the end of this century. Scientists are certain that human activities are changing the composition of the atmosphere, and that increasing the concentration of greenhouse gases will change the planet's climate. But they are not sure by how much it will change, at what rate it will change, or what the exact effects will be."[2]

The following statistics from the EPA illustrate the impact that construction has on resources in the United States:

- 70% of all electricity consumed is in buildings.
- 40% of all energy consumed is in buildings.
- 12% of all potable water used is in construction.
- 30% of all raw materials used are in construction.
- 30% of all carbon dioxide emissions are caused by construction.
- 65% of all waste output is caused by construction.[3]

Buildings and construction are integral to energy usage and fundamental to environmental damage. The damage is more than initial construction; it includes long-term effects as well. For example: How much repair will a building need over its lifetime? Besides construction, the cost of maintaining a house over its lifetime is significant. Repairs, energy usage, any necessary upgrades, as well as general maintenance can easily exceed the original energy cost of a home. Home repair and maintenance are directly proportional to the durability of the structure. The durability and quality of materials used and how they are installed influence the frequency and costs of their repairs over the life of the house. Some innovations in building materials initially have a poor record of durability or have a similar difficulty. Examples include the preservative treatments for exposed wood; fire-retardant, treated plywood; below-standard siding; aluminum wiring in branch circuits that do not have the proper connectors; and polybutylene pipes that do not have proper installation methods. Whereas some manufacturers improve their products early in their use, many ignore failures and instead change their product line or business plan altogether. Some of a product's inadequacies may not be discovered until several years later well past the warranty period. This unanticipated obsolescence adds to the hidden costs, that is, environmental damage, because the decayed building material will now occupy valuable real estate and reside in a sanitary landfill. Necessary repairs like this add unnecessary costs to a home that could have served a better purpose.

Energy Usage in Buildings

Over a home's lifetime, energy usage plays a very significant role in its efficiency and productivity. According to the United States Department of Health and Human Services, in 2003 the average cost of energy in homes was $1,527 per year. According to the Department of Energy,

[1]*Courtesy of NOAA, http://lwf.ncdc.noaa.gov/*
[2]*Courtesy of EPA, http://www.epa.gov/*
[3]*Ibid.*

in 2007 the cost of all energy used in an average home was $2,100, around half of which was to heat or cool the home. There are several reasons for the high cost, such as the elevated cost of all energy (gas, electricity, and oil), overuse or waste of energy (keeping lights on), unnecessary loss of energy through exfiltration or infiltration, and climatic changes. Consider that most homes last between 50 and 100 years without significant repairs. Accounting for a modest inflation rate of 7%, one can expect to spend up to $210,000 to heat and cool a home and operate equipment such as lights and cooking facilities. Some experts suggest a savings of up to 50% or even more on the energy bill with easy-to-use mitigation measures. Imagine saving just 25% of that cost? The financial savings alone may be worth the initial investment to curb energy usage. In addition, reducing the use of energy derived from nonrenewable natural resources adds to the savings.

Similarly, all these energy costs reflect harm brought to the environment in several ways. Directly, the environment is affected by the by-products of manufacture and the delivery of these building materials. The depletion of nonrenewable resources is essentially depriving future generations of these resources. Without proper management and a considerate approach toward the methods of construction, we are spending our children's inheritance—these same nonrenewable natural resources.

ENTER: THE GREEN HOME

Green building practices have evolved to address these concerns of energy usage and its detrimental effects on the environment. Building a *green home* can and likely will have elements of a sustainable design. However, a green home and a sustainable design need defining.

What Is a Green Home?

There is no strict definition for what makes a home green. *Merriam-Webster's Collegiate® Dictionary* defines the term *green* as "concerned with or supporting environmentalism." The United States Green Building Council (USGBC) is a nonprofit organization composed of leaders from every sector of the building industry who are working to promote buildings that are environmentally responsible, profitable, and healthy places in which to live and work. The USGBC has a simple, one-line definition of a green home: "A green home uses less energy, water, and natural resources; creates less waste; and is healthier for the people living inside."[4] Green homes can look like any other home. Figures 1-1, 1-2, and 1-3 illustrate different views of a green home.

The Southface Organization is a nonprofit organization whose mission is to promote sustainable homes, workplaces, and communities through education, research, advocacy, and technical assistance. This organization introduces its Earthcraft House as a green building program that serves as a blueprint for healthy, comfortable homes that reduce utility bills and protect the environment. In simple terms, green building is the thoughtful consideration of both indoor and outdoor environments during the design, construction, maintenance, and use of a building. Some environmental considerations include site-specific positioning, use of sustainable and renewable materials, application of energy-efficient building techniques, water conservation, indoor environmental air quality, and self-generation of energy. There is

[4]*U.S. Green Building Council (USGBC)*

Courtesy of Miller Custom Homes

Figure 1-1 ■ Green homes can look like any other modern home and still have features that are energy conscious and that are built with sustainable materials.

Courtesy of Miller Custom Homes

Figure 1-2 ■ Green homes can have attractive landscaping that harmonizes with the neighborhood yet have features such as permeable paving, xeriscape, or natural landscaping and reuse of gray water systems.

Courtesy of Miller Custom Homes

Figure 1-3 ■ The use of windows on south-facing walls allows solar heat gain during the heating season.

no established threshold that defines a home as truly green, however. Any contribution toward one or more aspects of environmental consideration or that leads to an increased conservation of natural resources could garner the term *green home*. The proponents for this movement are from different backgrounds and have different perspectives on the purpose and objective of a green home and, therefore, the solution. Enthusiasts of green buildings are no longer just the counterculture types. Green building is a mainstream practice as evidenced by the Green Home Building Initiative of the National Association of Home Builders (NAHB).

Consumers drive the demand for green building, so it is important to follow a plan or design that will truly make a difference in limiting environmental impact or that will save homeowners significant money on their investment for going green.

Arguments for a Green Home

By simply selecting this book, you have asserted that you agree with the concept of green homes but may need some motivation and encouragement to employ such practices in your home construction. Some reasons for making the decision to build a green home include:

1. Economic savings—There are definitely monetary savings realized in conserving energy and reducing water usage. More energy savings yield more monetary savings. Certainly, there are additional start-up costs apart from building with conventional materials. There are savings realized from conservative practices in energy savings, material use, or waste that easily compensate, and the return on investment is relatively short for most measures. Additionally, the economic savings are far from just personal. This is a global effort because it is a global problem. Although the solutions may seem to address personal financial rewards in energy cost savings, they can bring about improvement in the global economy as well as our way of life.

2. Environmental savings—The number one reason to build a green home is to participate in being kind to the Earth. Limiting your impact on the environment may seem like a small thing, and most people avoid this sort of contribution because their efforts seem inconsequential, but momentum builds on itself. The more you do, the more others around you will do until a critical mass is reached.
3. Societal well-being—The most common reason posited by those who choose green building techniques or methods is consideration of their children's future. Everyone is responsible for their actions. Our stewardship on this planet should be to protect our natural environment. The goal should be to reduce the rate of impact on the world environment.

WHAT IS SUSTAINABLE DESIGN?

A sustainable architectural design results from designing a structure to comply with the principles of economic, social, and ecological interests.

The U.S. Government Services Administration (GSA) explains that:

> *. . . sustainable design seeks to reduce negative impacts on the environment, and the health and comfort of building occupants, thereby improving building performance. The basic objectives of sustainability are to reduce consumption of non-renewable resources, minimize waste, and create healthy, productive environments.*
>
> *Sustainable design principles include the ability to:*
>
> - *optimize site potential;*
> - *minimize non-renewable energy consumption;*
> - *use environmentally preferable product;*
> - *protect and conserve water;*
> - *enhance indoor environmental quality;*
> - *optimize operational and maintenance practices.*
>
> *Utilizing a sustainable design philosophy encourages decisions at each phase of the design process that will reduce negative impacts on the environment and on the health of the occupants without compromising the bottom line. It is an integrated, holistic approach that encourages compromise and tradeoffs. Such an integrated approach positively impacts all phases of a building's life cycle, including its design, construction, operation, and decommissioning.*[5]

The U.S. National Park Service asserts that:

> *Sustainability does not require a loss in the quality of life, but does require a change in mind-set, a change in values toward less consumptive lifestyles. These changes must embrace global interdependence, environmental stewardship, social responsibility, and economic viability. Sustainable design must use an alternative approach to traditional design that incorporates these changes in mind-set. The new design approach must recognize the impacts of every design choice on the natural and cultural resources of the local, regional, and global environments.*[6]

[5] *Courtesy of the U.S. General Services Administration.*
[6] *Courtesy of the National Park Service, Guiding Principles of Sustainable Design (1993).*

A model of the new design principles necessary for sustainability is exemplified by the Hannover Principles or Bill of Rights for the Planet, developed by William McDonough, Architects for EXPO 2000, held in Hanover, Germany.

1. *Insist on the right of humanity and nature to co-exist in a healthy, supportive, diverse and sustainable condition.*
2. *Recognize interdependence. The elements of human design interact with and depend upon the natural world, with broad and diverse implications at every scale. Expand design considerations to recognize even distant effects.*
3. *Respect relationships between spirit and matter. Consider all aspects of human settlement, including community, dwelling, industry and trade, in terms of existing and evolving connections between spiritual and material consciousness.*
4. *Accept responsibility for the consequences of design decisions upon human well-being, the viability of natural systems and their right to co-exist.*
5. *Create safe objects of long-term value. Do not burden future generations with requirements for maintenance or vigilant administration of potential dangers due to the careless creations of products, processes or standards.*
6. *Eliminate the concept of waste. Evaluate and optimize the full life cycle of products and processes to approach the state of natural systems, in which there is no waste.*
7. *Rely on natural energy flows. Human designs should, like the living world, derive their creative forces from perpetual solar income. Incorporate this energy efficiently and safely for responsible use.*
8. *Understand the limitations of design. No human creation lasts forever and design does not solve all problems. Those who create and plan should practice humility in the face of nature. Treat nature as a model and mentor, not an inconvenience to be evaded or controlled.*
9. *Seek constant improvements by sharing knowledge. Encourage direct and open communication between colleagues, patrons, manufacturers and users to link long-term sustainable considerations with ethical responsibility and to reestablish the integral relationship between natural processes and human activity.*[7]

Sustainable Design and the Life Cycle of a Building

Integral to the understanding of sustainable design are the phases in the life of a building. Just like a living entity, a building is conceptualized, is created, lives a productive life, and then dies. Just like our lives, each phase has variations. For example, the ending of a building's life may be gradual due to natural decay, or it may be dramatic as the result of demolition or even a natural event such as an earthquake, a tornado, or a hurricane.

There are four major phases in the life cycle of a building: prebuilding or design phase, construction phase, useful life phase, and postbuilding phase. During each phase, special considerations mark a sustainable design. The first phase is marked by site selection, building design, material selection, and construction method. The second phase, construction, is concerned with the assembly of all the parts together. A green construction project is concerned

[7]*Courtesy of William A. McDonough.*

with minimizing site impact, recycling, and processes applying design parameters. The third phase is useful life and includes the building's operation, maintenance, and repair. How long it is used relates to its durability. When a building has completed its useful life as intended, it is deconstructed or moved and reused. A green home that is deconstructed is either reused or its material components are used in recycling. This marks the last phase and leads to a new cycle.

Sustainable design and materials fit into this design by applying conservation principles. These four major stages in the life of a building have more depth that is important to understand when considering a green home. For instance, during the prebuilding or design stage, one considers the selection of building materials and the components of the building. In addition, material selection is more than just the direct costs associated with manufacturing the material or equipment used for construction. It involves the cost of *embodied energy*, that is, the total amount of energy associated with the product. It would include the acquisition of raw material, mining, transportation, processing, manufacture, further delivery to retail outlets, further delivery to the site, and installation. Some products have considerably more embodied energy than others. For instance, if two similar products are available from different suppliers and the cost difference is measured in pennies, the initial decision may be based on cost. But if one product must be transported a thousand miles farther, the energy cost and environmental impact, although not evident to you, the final consumer, will be felt with the additional, unnecessary expenditure of oil and gasoline in transportation, degradation of the highway system (and essential repair), added environmental impact of the spent energy, and other comparable costs.

SHADES OF GREEN

The commitment to building green will be marked by the degree of your participation. This depth is characterized by the abstract *shades of green*. The more you invest in elements or technology, the darker your shade of green will be. Some professional associations offer relative values between various green features to help assist you with a comparative value for different aspects or measures. The USGBC discusses shades of green.

> *The phrase "shades of green" is often used to refer to various levels of achievement in adopting resource efficiency in a home. Homes with one or two green measures are sometimes called light green while homes with several green measures are called dark green.*[8]

The USGBC promulgates the Leadership in Energy and Environmental Design (LEED) standard for evaluating the contribution made toward environmentally responsible building materials and practices. The LEED Green Building Rating System™ is one nationally accepted benchmark for the design, construction, and operation of high-performance green buildings. There are four levels (shades) of green developed as a rating system by LEED: Certified, Silver, Gold, and Platinum; with Platinum being the highest level of LEED certification.

According to the USGBC:

> *LEED homes are safer, healthier, more comfortable, and more durable than conventional homes. The benefits of a LEED home include economic benefits such as lower energy and water bills; environmental benefits like reduced greenhouse gas emissions; and health benefits such as reduced exposure to mold, mildew and other indoor toxins. Even better, the net cost of owning a LEED home is comparable to that of owning a conventional home.*[9]

[8] *Courtesy of U.S. Green Building Council (USGBC).*
[9] *Ibid.*

These shades of green allow you to quantify your participation in green building. The design stage is where your shade of green will be determined. Design decisions must include highly effective and cost-efficient strategies to eliminate design problems of the building. This is easier to do at this stage than later when the building is in the conceptual stage. There will be conflicts between design approaches and a hierarchy of strategies related to optimizing the design that is worked out at this stage. For instance, say you want to install a passive solar water heating system. The added weight of water directly affects the necessary strength of the building, calling for stronger bearing members. Increasing the size and amount of wood to serve as support makes less environmental sense than considering a renewable or recyclable alternative such as steel. There will be many design considerations like this that call for mitigation of effects that still allow for the intent: a truly green home.

Examples of Green Homes: Recycling to Earthship

There are many variants of construction that claim the term *green building*. They range from simple household practices to utility-based independence. You may find your choice for investment to fall somewhere in between. A particular shade of green portrays the depth of your participation in green building.

Recycling, Changing Bulbs, and Choosing Alternative Power Sources

Almost all cities and towns offer recycling options for materials that are easily reused such as cardboard, paper, plastic, and glass. These items have a triangular arrow symbol that indicates they are recyclable. This symbol is called the Möbius Transform. Shown in Figure 1-4, this symbol has grown to define recycling activity. Recycling or reusing construction materials

Courtesy of iStock Photo

Figure 1-4 ■ The Möbius strip or Möbius band is a surface with only one side and one boundary component. It has the mathematical property of being nonorientable. It is also a ruled surface. It was codiscovered independently by the German mathematicians August Ferdinand Möbius and Johann Benedict Listing in 1858. It has become symbolic of the recycling effort.

accounts for the least amount of embodied energy. Instead of hauling construction materials to a landfill and eliminating precious raw land, reuse of the materials is the simplest way of reducing environmental impact. You can make a difference by changing an incandescent bulb to a fluorescent type. The savings accumulate over time and such changes are simple.

Supporting your power company's use of alternative energy is another way to be green. Some energy companies are now offering an option to subscribe to alternative energy production sources. It elicits a slightly higher fee per unit of power; but, essentially, it allows the power company to support these alternative power options such as wind, solar, or geothermal. This is a simple yet meaningful step toward being green.

Modest Green Design Measures

Adding simple design elements or low-impact materials to your homebuilding project is a more meaningful way of contributing toward a green home. For instance:

- Considering sites such as a brownfield or grayfield, thus making use of existing developed real estate instead of eliminating undeveloped land
- Designing with rainfall runoff in mind and avoiding too much pavement
- Building the minimum size home instead of a mansion
- Coordinating the construction project to allow for quick transition to habitable conditions

The sequence of construction activities is arranged based on the materials, architecture, and structure. There are only so many ways that a house can come together based on its design. Length of construction activity could adversely affect construction cost and even environmental impact. The more time it takes to build a home, the greater the costs will be with interest, delays, and so on. A general contractor's main job is to coordinate laborers, trade subcontractors, and material suppliers. Knowing how long the foundation will take; how long the carpenters need to assemble the walls, floor, and roof; when to get the electrician in (and out); when the interior walls are ready for painting; and similar questions affect their effort to be efficient and Earth friendly. Coordination among subcontractors and material suppliers is necessary to reach the quickest completion of each trade or installation. An owner-builder supervising the work can succeed by discussing the time needs with the subcontractor and asking the right questions.

More Extensive Green Design Measures

During the design process, you set the pattern for how you will build your new home. The more thought you give to the process at this point, the better the achieved design. Your home is the assembly of a wide variety of materials and numerous people who will assemble the materials; some people have never met you before and are not acquainted with your expectations for quality or installation standards. It will be your job to convey those expectations to those tradesmen and laborers. You can do that in several ways.

The construction document and plan you develop in the design stage function as the tools to communicate your intent for quality, durability, and even a green building. Construction plans like those shown in Figure 1-5 are commonly called blueprints. These blueprints are just blue-line copies of the original plans drawn by an architect. The subcontractor will base an

Courtesy of iStock Photo

Figure 1-5 ■ A good design is where a green home begins. Systems and interrelationships between materials and technology are resolved at this point.

estimate of the work on a review of your plan and specifications. Construction documents are considered to be an extension of the contract between parties. The plans and specifications are the best way to set out the degree of quality you intend for your green home. The type of materials and their quality must be specified on these documents. You can even specify the brand name and model number for specific items such as appliances, carpets, doors, and windows. This will help ensure that what you want is what you will get.

A higher level of commitment is on-site energy production that supplements the electricity and fuel consumption in the home. Wind, solar, water, and geothermal are sources of free, renewable energy that are available to us. Investment in the equipment to harness and use this energy is significant but still has a cost-return cycle that makes it affordable.

For both you, the owner-builder, and the general contractor, the single most significant measure you can take to ensure a quality design that will meet your goals while considering the architectural design and meeting every other legal requirement is to hire a registered architect who is well versed in sustainable design. Most architects are familiar with green building and know how to guide owner-builders through the maze of design considerations and legal hurdles that they will likely face, including local zoning requirements, subdivision restrictions, and similar building limitations. Another reason for selecting an architect is that he or she knows the Building Code and how your project will conform to those regulations.

Premier Green Measures

At the extreme end of the green home spectrum is the self-sustaining Earthship, a truly autonomous building. Such a building is designed to be operated independently from any support services of public utilities such as electricity; water supply; sewerage; gas storm drains;

communications; television; and, in some cases, even solid waste disposal (usually involving recycling). Advocates of Earthship construction describe advantages that include zero energy costs, minimal environmental impacts, and increased self-sufficiency. These Earthship homes often rely very little on municipal or public services and are not adversely affected by power outages. Generally, they are off the electric power grid. Proponents of the Earthship are normally very independent and self-sufficient. They garner potable water by harvesting rain in a cistern. They reuse gray water, and some even salvage and use human waste as compost in gardening. The Earthship is normally an earthen structure but it can be built out of waste products such as vehicle tires filled with earth, which many times are built into the side of a hill to retard energy loss from exfiltration. It is more of a lifestyle choice than normal contributions made toward the environment. This type of house is at the extreme end of the green building movement and will probably not meet the housing demand for most of the nation.

Durability and Quality Control as Shades of Green

Another level in the design process is reached by building a home while considering quality control and material durability. To achieve a darker shade of green home, the longevity of the home must be considered. One must consider the structural ability of the home to survive natural forces such as wind or seismic events or even greater natural disasters such as a hurricane. The life of the home must also be considered. The longer a home lasts, the better it is for the environment because less waste disposal occurs from discarded building materials and fewer materials will be needed for its replacement. After the design stage, a quality control program that closely observes the work done at your home will ensure that the quality of the products is exactly what you have specified and what is being used instead of a similar, cheaper version or something that is not ecofriendly. Subcontractors can be bargain hunters when they see an opportunity and may believe it does not matter to owners. Additionally, the manner in which materials are assembled is critical to their functionality. To meet your green design, the construction documents including a plan might specify that a wooden wall frame use minimal lumber with 24-inch spacing of studs and a single top plate or maybe two-stud corners. This requires the addition of bracing and connectors, which is something not ordinarily done. It is important to convey the importance of following your design. Some of these workmanship issues are not part of the inspector's authority to regulate and unless you catch it, that substandard work will be part of your new home. It will be up to you to ensure this higher standard.

As an owner-builder, you can set up a quality management program and oversee this work performed by subcontractors. You can perform this quality control yourself or you can hire others. If you have considerable experience in construction, this is an avenue for cost savings during the building process. If you have limited skills, you should hire someone to represent your interests. Your learning curve could take longer than this one construction project and your perception of mistakes could be seen more as interference than an oversight and cost you in other ways. The quality control should be set up as part of the contract, letting the subcontractor know ahead of time that the quality of the work would be overseen and must be approved before the next sequence of work could begin or, more importantly, before the next payment could be issued. The first real step toward sustainable and green building design is the design decisions made at this preliminary stage. Shades of green; cost; size; location; site considerations; proximity to schools, shopping, or work; and the other myriad of choices come together to help you along the path of green building.

FIRST REAL STEPS—PLANNING AND DESIGN

Energy efficiency, sustainable materials, durability, and quality assurance are a few of the considerations toward your goal of an environmentally conscious construction project. However, the first and most important pragmatic step is the design. So many building system problems develop from poor planning or haphazard design. In construction, when you decide to cross that bridge when you come to it, the results usually look like the fix they are as opposed to a well-thought-out plan. Design decisions are important, and cost-effective strategies are the result of forward thinking by designing problems out of the building when it is on paper or in your head.

Site Selection and Design

The design should still meet the conditions of the proposed site. In some ways, site selection is equally important unless you expect to find the perfect lot for your design the first time. The design for your home must fit the numerous parameters offered by the site. Where you build your dream home will be a decision based on many factors, some personal and others technical. You will want your home to fit into the neighborhood and still be where you want to live. Although a green home can look like any other, it could also use south-facing windows, a xeriscape (a landscaping method that uses drought-resistant plants in an effort to conserve water), gray water systems, water harvesting, earthworks, wind generators, a ground source heat pump, and solar thermal or photovoltaic panels. This strategy would certainly stand out in most traditional neighborhoods, but there are other factors besides aesthetics. You will want to have minimal effect on the land you develop.

Less development in the form of backhoes and earthmoving equipment brings less environmental impact and reduced costs to your development. You must think in terms of using the natural terrain to your advantage as much as possible. Remember the importance of the design considerations in relation to the site conditions. If the building site has a natural slope, the house can be designed to utilize this feature. If creating a three-sided basement, forming an earthen barrier, remember that the two stories you have planned need to remain within the city zoning height limits. Terrain should allow for natural flow of rain drainage around or away from your building site as well as naturally flowing to where you want it and away from where you do not want it. Otherwise, you will have to accommodate nature by adding drain pipes and landscaping. Also consider if your site allows for creation of a suitable microclimate. These can be formed by elements of your home such as fences, trees, shrubbery, water features, and even paved surfaces to create subtle but very real differences in temperature and humidity around your home. For example, placing trees and shrubbery in strategic spots on the lot could provide heating and cooling benefits.

If the lot offers good southern exposure, there could be a good investment in active or passive solar potential to generate heat and energy. In the cooling season, natural conditions that mitigate heat from the sun such as deciduous trees or flowing water would be beneficial. There are lots of considerations and compromises that will affect your plans to build a green home.

Architectural Design

The term *architecture* includes many elements that make a house feel more like a home. These elements include size, shape, style, circulation and usability, lighting, and structural design.

Selecting the right size is a significant first step. Too large of a home is a waste of resources. Too small of a home is unusable and becomes a discarded project. Style and house shape are important to a green home. Keep the design small and simple yet thoughtful. Plan for the future and consider phased construction. You have to consider a harmony of elements and circulation patterns. Lighting plays a part in a green home. Too much artificial light becomes an energy drain in lumen cost and loss of conditioned air if fenestration is intended for natural lighting. Another natural resource that green homes harvest is natural daylight by channeling light to interior spaces. Where you locate sleeping rooms, the laundry, and bathrooms will improve the ease of use of your plumbing facilities. Also, clustering water facilities such as the kitchen and bathrooms saves on plumbing materials and water usage. Consider the distance between the hot water fixtures and the water heater. A shorter distance saves on water and energy. The length of duct that runs from the heating, ventilating, and air-conditioning (HVAC) equipment is inversely proportional to heating and cooling efficiency and directly proportional to energy loss.

Another architectural consideration is ceiling height. The Building Code has a minimum height limit for habitable rooms but there is no maximum. Don't exceed the minimum height too much. Remember, heated air rises and cooled air falls. Heat flows from warm to cold. The volume of air necessary to condition (heat or cool) increases proportionally to every vertical increase in ceiling height. You should also consider the benefits and drawbacks to windows. Windows add natural light as well as a beautiful view inside your home. They also present a source of heat loss, allowing conditioned air to escape and outside air to infiltrate (when open), bringing moisture, dust, and allergens inside. Moisture can cause problems for any construction material and promotes the growth of mold and other health hazards. Moisture flows from damp to dry. Insulated glazing will compensate for what otherwise would be an energy loss while still providing the desirable view. All these things are part of the architectural design process. This is where you build the quality for your project: by thinking ahead. There are trade-offs and compromises when building any house. A green home presents some peculiar challenges and calls for trade-offs that sometimes seem bewildering.

Design Consideration for Materials and Techniques

The selection of materials and deciding how they are installed are significant considerations in making your house a fully green home. Not only do materials affect the energy efficiency of your new home, they define the degree of impact on the environment after their useful life. Of course, there is another set of trade-offs or decisions to make with material selection and that includes whether or not you will do the work yourself and your experience level with nontraditional building materials that may have a good reputation for energy efficiency.

Remember that the objective is to have a successful project that is also environmentally responsible. Therefore, selecting a material also includes ease of installation and your familiarity with the product. Material selection may be affected by these and several other criteria. For example, although concrete is less desirable from a sustainable perspective, it is quite durable and is quite energy efficient if installed within the confines of an insulated concrete form (ICF). You might want to have a passive solar design with thermal mass. Concrete makes an excellent candidate for this, yet its production creates by-products (e.g., sulfur dioxide) that are harmful to the environment. Although concrete is durable, its long-term impact still adversely affects the environment. However, the corresponding reduction in energy consumption may offset this impact. This type of trade-off is what green building is all about: control of energy usage, limitation of environmental damage, cost savings, and the use of durable materials to avoid premature use of the landfill.

Design for Energy Production and Power Generation

The first step toward energy production is the need to optimize the performance of the thermal envelope and mechanical systems. Doing so reduces the need to produce that amount of energy. The next step is adding a renewable energy system. You can generate your own power; supplement your energy needs; or, at the extreme end, even unplug from the utility companies. Electrical energy at a domestic installation is most commonly generated from solar, wind, or water power. Which type and how much are up to you. In some cases you are limited only by the local zoning ordinance. You can have a roof full of photovoltaic arrays, a wind turbine generator like the ones shown in Figure 1-6, and even a hydroelectric turbine

Courtesy of iStock Photo

Figure 1-6 ■ A green home may even include equipment that is able to generate electric power for private use. In many cases, any residual or excess power can be sold back to the main service provider.

generator if running water is available. In some conditions, their use is regulated by other state and federal agencies that protect the environment in other ways. However, being off the grid has its own set of challenges. To maintain a regular flow of electricity when the sun is not shining or the wind is not blowing, these systems use batteries. Most are expensive and have short life spans and must be maintained. This requires a bit more work than most homeowners care to invest.

Water Conservation through Efficiency and Design

Saving energy is only one part of an environmentally responsible home. Another factor to be aware of is how well a home conserves other natural resources like water. The design should be evaluated for the most efficient use of a freshwater supply. Landscaping, if done properly, will help conserve energy (by dissipating heat during summer and conserving heat during winter). However, there is a cost to this landscaping: water, which is another valuable natural resource. You can plan landscaping that consumes less water. Xeriscaping could offer the option that benefits your objective. Reusing water through gray water systems is time tested and is proven to work. These systems reuse the drain water from the shower, bathtub, and sinks to irrigate the landscaping or garden. This gray water is sometimes used to flush the toilet as well. Rainwater collection for reuse is known as water harvesting. In an average home, we use about 100 gallons of water per person per day. For a family of four, that is about 150,000 gallons per year, or 11 million gallons over a lifetime.

Saving half that amount of water with harvesting makes perfect sense. Figure 1-7 shows how rainwater harvesting should be planned at the design stage to avoid conflicts with other plumbing installations. The cost of potable water includes the cost of delivery, which includes energy for pumping. Flushing toilets, washing cars, and watering landscaping are just a few of the uses for harvested rainwater. Cisterns have made a comeback, especially in dry, arid climates where water is a precious commodity. Although a great idea, there are some local issues that may affect your plans to harvest water for reuse. Some localities have restrictions on water harvesting because water rights for legal entities downstream prevent them. Therefore, whatever the plans you make, research the affecting law.

Sustainable Materials in Green Homes

The elements of sustainable design and green building are discussed further later in this book, but the following few examples are considered based on the logical steps in the construction process:

- Site work and landscaping
  - Low-impact development such as carefully locating buildings, minimizing impervious surfaces, and storm water infiltration. These measures help to preserve wildlife habitat and prevent erosion.
- Footing and foundation
  - Concrete block or frost-protected shallow foundation instead of poured-in-place concrete generally lessen material usage.

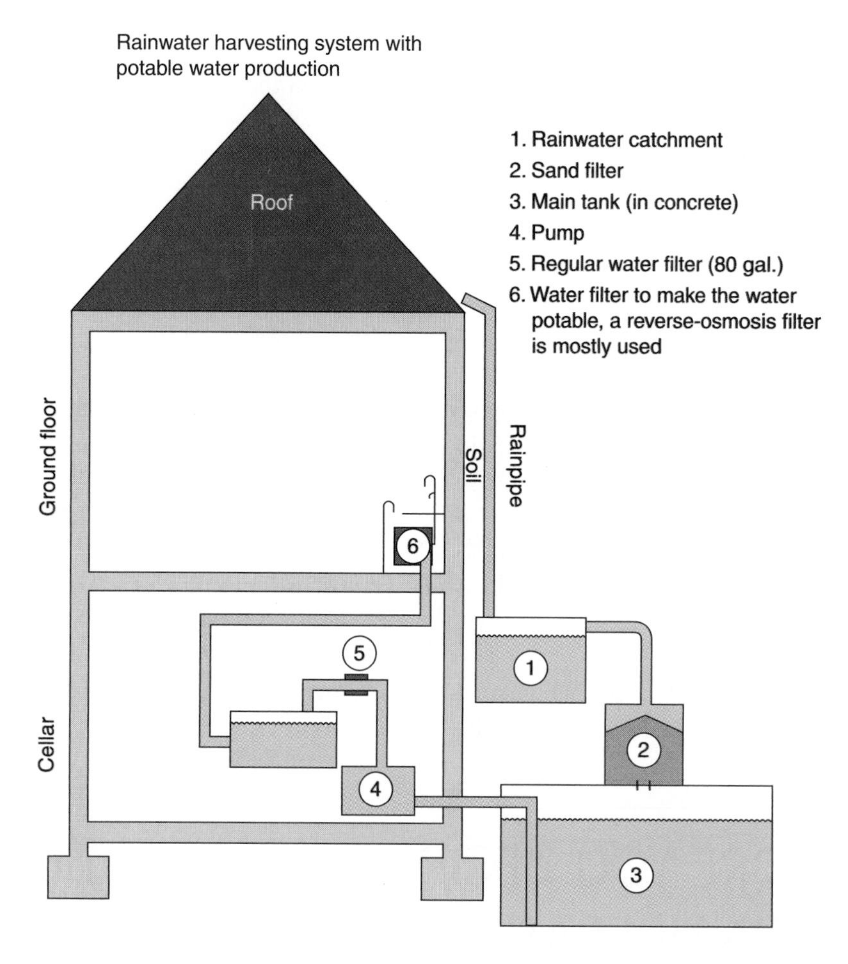

Delmar/Cengage Learning

Figure 1-7 ■ A green home may even include a mechanism that will harvest water for private use. Rainwater is generally pure and potable. Its reuse is part of a green home.

- Structural framing
  - Adobe block or earthen walls represent a truly renewable material resource.
  - Straw bale wall framing is a low-tech way of achieving the intent of sustainable design or construction while still providing energy conservation measures.
  - Alternative manufactured materials such as building panels or engineered trusses provide increased strength but use fewer materials.
- Building envelope
  - The use of stucco, straw paneling, or fiber siding has advantages to traditional wood or synthetic siding.
- Energy efficiency
  - Air tightness: 25% to 40% of the total heat loss or gain in most houses is associated with air leakage.
  - Superior insulation retards heat flow.
  - Better-quality glazing and doors provide low emissivity and higher thermal performance.

- Roofing
  - More durable, longer-lasting materials can lengthen the life span of a building.
- Interior finishes
  - Paint with natural pigments, natural wall coverings, and wood veneers
- Flooring
  - Natural materials for carpet, reuse of flooring materials, cork material, or long-lasting material such as ceramic or porcelain tile.
- Plumbing
  - Water conservation fixtures, gray water systems, and composting toilets; use of rain-water harvestings; solar water heating.
- Mechanical
  - High-efficiency heating and cooling systems.
  - Energy conservation and heat exhausting systems; use of ventilators.
- Electrical
  - Multiwire branch circuiting practices reduce material. Using alternative energy sources such as photovoltaic, wind, or hydroelectric.

BEFORE YOU DECIDE . . . REFLECTIONS AND CONSIDERATIONS

✔ Although building homes is a necessary component of society, a green design recognizes that construction of any type adversely affects the natural environment.

- Wasteful practices during construction unnecessarily fill landfills.
- Natural terrain conditions that are capriciously distorted to match the home design lead to long-term environmental damage.
- Overuse of potable water is both a resource loss and an energy loss.
- Options include conscious decisions to design and manage construction projects.

✔ Green homes are those that provide adequate shelter and meet the needs of the occupant while reducing environmental impact with thoughtful consideration for material choices and use of energy-efficient equipment and methods of construction.

✔ Sustainable design is the selection of materials for construction that include renewable materials or that cause minimal environmental impact while improving the overall health of the occupants. A sustainable design is one that achieves independence from the use of nonrenewable natural materials.

- Thoughtful consideration during the design process can significantly reduce environmental impact and concurrently reduce a lifetime cost of energy.
  - The size of the project should reflect the need, not the prevailing norm.
  - Energy-efficient design decreases energy demand.

    - Alternative energy production uses free natural resources such as the sun, wind, geothermal, water, and others.
    - Use of renewable building materials instead of nonrenewable natural resources promotes a sustainable building practice.

- ✔ Building life cycle illuminates the elements of land use over its lifetime. The four stages of construction planning include:
    - Preconstruction
    - Construction
    - Useful life
    - Deconstruction (removing and reusing)
- ✔ Embodied energy reflects the overall environmental footprint of a material or product irrespective of the apparent value or direct cost.
    - Ignoring these hidden costs causes environmental damage elsewhere.
- ✔ The measure of participation in green building practices reflects the shade of green for the project. Modest measures are lighter shades of green. More serious measures represent darker shades of green.
- ✔ The first real step toward a green home must begin with proper planning and a design that includes a holistic approach to providing shelter within the confines of a fragile natural environment. This design must include:
    - Site selection
    - Architectural design
    - Material selection and methods of construction
    - Alternative energy production
    - Water conservation

FOR MORE INFORMATION

Department of Energy
http://www.eia.doe.gov/

Environmental Protection Agency (EPA)
http://www.epa.gov/climatechange/basicinfo.html/

The International Code Council Evaluation Service SAVE™ site
http://www.icc-es.org/

National Association of Home Builders (NAHB)
http://www.nahb.org/

National Oceanic and Atmospheric Administration (NOAA)
http://lwf.ncdc.noaa.gov/

National Park Service
Guiding Principles of Sustainable Design, 1993
http://www.nps.gov/

Southface/Earthcraft House
http://www.southface.org/

United States Department of Health and Human Services
http://www.acf.hhs.gov/

United States General Services Administration
http://www.gsa.gov/

United States Green Building Council (USGBC)
http://www.greenhomeguide.org/what_makes_a_green_home/green_homes_101.html/

William A. McDonough
http://www.mcdonough.com/

Chapter 2

GREEN HOME SITE SELECTION

GREEN HOME LOT SHOPPING

The first step in building a green home is to determine the location. Careful planning is necessary during this step. There are three important principles to consider regarding the value of land: location, condition, and usability. Take your time shopping for a lot. Consider everything that affects your plan to make a green home. The building site is critical to many aspects of what makes a home green. A bare lot like the one shown in Figure 2-1 may be an option you select. Other choices are available, such as a previously developed lot.

Courtesy of iStock Photo

Figure 2-1 ■ Site selection is based on numerous factors, including its proximity to traditional services and the ability to get a permit based on Zoning and Building Code compliance.

Proximity to Services

First, consider the distance from the lot to such facilities as the grocery store, schools, medical care, shopping, and entertainment. The distance you travel each day will add to your impact on the environment. Take into account how central the lot is to the normal activities you engage in from day to day such as how much you will have to use a vehicle to drive to and from work and how often friends and family come to visit. The cost associated with fossil fuel use *after* your home is complete could be significantly greater than the direct costs of the home itself. The effects of spent fossil fuel present themselves as smog, greenhouse gases, accumulation of solid waste, damage to crops, and increased atmospheric carbon dioxide that could lead to global warming. Delivery of fossil fuel has led to oil spills that further damage the environment.

You could reduce your vehicular travel if the site is close enough to bicycle to work. Living as close as is reasonable to these conveniences that you use regularly will add to the benefits of building a green house. A true aspect of a green home includes planning for less fuel-powered vehicular travel. A central tenet in green homes is to think of generations ahead. Imagine the direct and indirect costs associated with having a home at this site over time. This single criterion will have an enormous impact on your shade of green because the effects will be felt for a long time.

Location and Neighborhood

The building location is almost as important as the design itself. Even if the house is the perfect design and has every amenity, it could be next door to the most offensive conditions imaginable. Additionally, it may be of importance to site the house with a particular exposure to the south (or east, west, or north) in order to capitalize on natural conditions such as solar exposure of convective wind currents. Consider neighbors as well. Remember, your project will affect them until it is complete. The school your children attend will shape their lives. The wind and dust patterns prevalent in the area may cause cleaning to be a nightmare. Regard your goal of a happy, healthy life with the neighborhood or surroundings. When it comes time to sell, the marketability of your home will be directly related to the site location and condition.

Personal and Family Considerations

Your home should be a place where you feel comfortable and relaxed. The goal is to rest easy in the knowledge that your neighborhood is a community whose residents have similar hopes, dreams, and expectations. Many factors play into this feeling of belonging.

Lifestyle Consideration

An appropriate neighborhood should fit your family's needs and lifestyle. Imagine attending a get-together or barbeque with neighbors. Think about your children cutting a neighbor's lawn. The care of the neighborhood should match the care you would give to your home.

Overall Appearance

There are many aspects of green homes that are unseen. But there are some that may be visible and tend to stand out. Solar panels, wind generators, and rainfall cisterns are just a few of the features that could be an eyesore to your neighbors. The ones you select should be part of a neighborhood where those additions will be accepted by your community. Normally, such questions are addressed by the zoning ordinance in the city or county where you will build. In many cases, they can be found in the neighborhood association's rules and regulations.

Size

In the interest of environmental design, the smaller the house, the better. Yet, if it is too small, it may not fit into the neighborhood. It is equally important to think of your long-term economic investment while supporting the environment. A house that is either too large or too small relative to its neighbors may have resale problems. If it is too large, it will likely be the most expensive and suffer the greatest tax burden. If it is too small, it will seem like a poor investment compared to its neighboring houses.

Neighbors

Although difficult to determine for certain, it is very important to your short- and long-term goals to know if your neighbors will tolerate your green choices. You will have a difficult time during the construction stage if your neighbors are opposed to your plans. Complaints of noise, dust, debris, and garbage will likely be the preliminary signs that you have a problem. On any construction jobsite, construction debris and waste should be controlled constantly for a green home. Otherwise, you may have frequent visits from the police or zoning and building inspectors asking you to satisfy a complaint of this minor nature. When neighbors support you, you may learn about these trivial complaints with a friendly visit. Act on these complaints quickly to keep in their good graces.

Community Values

It is important to choose a neighborhood that upholds the same values and expectations that you do with regard to the community. Look at the infrastructure, such as streets, sidewalk, public facilities, trees, grass, traffic signs, and other indications about the pride of the community. The upkeep of the homes and the lawns should indicate the values of the neighborhood. A neighborhood such as the one depicted in Figure 2-2 may be exactly what matches your desires. Be sure to shop for a neighborhood that matches your values for a community.

Courtesy of Fotosearch

Figure 2-2 ■ Select a neighborhood that matches your lifestyle as you and your family grow older.

Around the Clock

Things change during the day and night in urban and suburban neighborhoods. Check out the traffic and other characteristics that could affect your enjoyment of your new home. See if a street comes alive with children playing in the street during the day, or if there are police patrolling in the locality. You could even visit the commercial district nearby at different hours of the day and see if anything surprises you.

Future Developments

Ownership of land is public record. Check with a city or county clerk to see who your future neighbors will be. Visit with them about your plans. Find out if there are future plans for land development near your lot like a large shopping center or junior high school. Check with the jurisdiction (Planning or Public Works Department) to see what future plans are being discussed.

Lot Coverage, Zoning Restrictions, and Other Legal Questions

Zoning restrictions may affect your design for a green building. It is almost always regulated by local government, that is, your city, town, or county. Local government regulates the physical development of land and the allowable uses for each grouping of properties. Enacted and enforced by a community, zoning ordinances establish regulations for the appearance and livability of a neighborhood. These laws typically restrict the region where conflicting uses such as residential, industrial, recreational, or commercial activities may take place. An example might be establishing a residential zone that allows only single-family detached homes instead of multifamily buildings such as apartment buildings or duplexes. In the case of a commercial zone, laws might regulate certain kinds of businesses such as an assembly use like a church within a residential housing subdivision. The zoning regulates public use in an area that includes nightclubs or bars. Among other things, zoning regulates the quality of life enjoyed (or endured) based on the level of appearance and condition of the property. In addition, some zoning ordinances regulate the size, height, proximity to property line (setbacks), building style, outside color, and architecture permitted in the specific "zone." Although there are challenges in gaining approval, these ultimately improve the long-term value of your investment. Restrictions imposed by your local zoning ordinance will affect your objective. Speaking with the zoning administrator to discuss your plans can prevent headaches later. The administrator can quickly tell you most of the land use controls that will be encountered if a certain lot is selected. When visiting with the zoning administrator, bring a copy of the survey or site plan with your house superimposed, along with a rough sketch of the planned design. Ask the difficult questions and see if your plan will be approved.

What Is the Allowable Lot Coverage?

Zoning regulations in each jurisdiction include the aspect of the amount of land that can be developed or covered. Development of the lot and its coverage normally include everything that covers the natural grade or restricts natural precipitation from draining into the soil. For instance, the building itself would certainly be considered when determining lot coverage. However, some other installations qualify components as lot coverage that may not be expected, including swimming pools; porches; decks; sheds; and, in some cases, fences.

Green homes are expected to improve on whatever the lot coverage is set out by the local jurisdiction. This is an important aspect of the green movement: to minimize your effect on the Earth. Less construction activity helps achieve this goal. An example would be a jurisdiction such as a beach resort that limits lot coverage to 30% and counts such items as second-story porches, outside stairs, driveways, and decks in that coverage.

The size of the house that is needed, along with the maximum lot coverage, will determine if the house you like will fit on the lot you are considering. Most people do not consider the building size until later. Keep the end result in mind when shopping for a lot. These are a few aspects to consider:

- Setbacks from property lines
- Maximum lot coverage
- Building height in stories and feet
- Building materials and aesthetics

What Are the Required Setbacks?

In land use jargon, a *setback* is the distance a building or other structure must be from the property line. Additionally, setbacks may be a distance away from a public street or road; an ascending or descending slope; or a waterway such as a river, stream, lake, or even a floodway. It could also be the distance away from a buried utility easement, utility meter, street, driveway, or swimming pool. Depending on the jurisdiction, other things like fences, landscaping, septic tanks, and various potential hazards or nuisances might be regulated. Setbacks are generally set in municipal ordinances or zoning law. Along city, town, county, state, provincial, or federal highways, setbacks may also be set in the laws of the locality, state or province, or the federal government.

A setback from the property line is implemented to prevent buildings from being too close together. This is a fire hazard as well as being uncomfortable for residents. Setbacks sometimes provide access for public utilities such as mailboxes and electric, water, sewer, or gas meters. A human corollary is that of personal space, that distance we give each other when we interact in a social or professional setting. Americans generally preserve about 18 inches as a bubble around them. Interacting with another, the distances are added together to predict about 36 inches between people. The common exception is in an elevator where it is socially acceptable for people to crowd together.

Most lots platted in the early twentieth century are narrow and have short setbacks. Because of this, older buildings may have a narrow separation between buildings. Distances of less than 5 feet are common in neighborhoods built in the United States before the turn of the century. Modern zoning principles established with guidance from past experience and professional associations have gradually increased this setback distance, thus reducing the buildable area while increasing the space between neighbors. There are setbacks from all sides of your lot: front, rear, and side yards. Each of these setbacks must be met for your building. You may own the property, but your right to build is restricted to the buildable area. This is in addition to the space reserved for setbacks. The buildable area in most residential districts is the entire lot, minus the setback requirements on all sides. A construction site plan normally provides an illustration of these front side yard, and rear yard setback requirements. Figure 2-3 illustrates a typical construction drawing that includes a site plan. These construction documents are required in order to apply for a permit. Normally, they are drawn by a registered architect. Because the

Courtesy of photos.com

Figure 2-3 ■ A typical site plan for residential construction will provide information on lot size, grade, and slope. It will illustrate intended setbacks, finished landscaping features, and access, as well as locations for all improvements (buildings).

zoning ordinances of various jurisdictions have changed over the years, many existing buildings lawfully intrude into these setbacks established for new development.

What Is the Maximum Lot Coverage?

Various zoning ordinances across jurisdictional boundaries, such as cities and towns, will have different density limits for the amount of the lot that can be covered with a building. Normal limits for lots within suburban subdivisions or developments are between 25% and 33%. In rare instances such as in highly urban settings, the density could be 100%, or covering the entire lot. Density represents the percentage of lot coverage allowed by the locality. It serves the purpose of aligning the appearance of the neighborhood.

What Is the Maximum Zoning Height?

Although the height of a building does not necessarily relate to a green building, complying with these ordinances is essential to be granted approval to build. Maximum height of any building is normally regulated by the local zoning ordinance and by the Building Code. The height, measured from the finished grade, is typically an average measure around the building or from the lowest level in case the finished grade varies in slope. The peak of the roof could be the upper limit for the measurement, or the ordinance could specify the average height of the roof line. This is normally measured as the median distance between the roof eave and the roof peak. However, some jurisdictions have alternative ways of establishing the reference datum, so some calculation may need to be done to make a determination. In any case, ask the local government how the height is measured in its jurisdiction.

What Are the Exterior Building Materials or Colors Required or Allowed?

There are some communities that require or prohibit certain materials or exterior colors to be used. For instance, a subdivision may require masonry to be used as exterior veneer. Conversely, there may be a prohibition against the use of certain materials such as shiny metals used on roofing or siding. Some subdivisions may not allow vinyl siding or exterior insulating finish systems (EIFS). These conditions may be established as zoning requirements enforced by the jurisdiction or by a homeowners' association as restrictive covenants. In any case, the materials you select for your home could be affected by these restrictions.

Building Code and the Building Lot

The Building Code sets out minimum regulations that ensure public safety and sanitation within buildings. The minimum standards such as setbacks are there to preserve public safety. Requirements like the minimum setback serve to limit the spread of fire in a conflagration.

Although flexible enough to allow green building design, the Building Code is restrictive in several ways that may affect building plans. For aspects of the site selection, the Building Code limits the size and placement on a lot in a similar manner to zoning. The setbacks from a property line are normally less restricted than zoning or any subdivision restrictions. In some cases, you can provide a rated wall assembly and be up to the property line. However, openings such as windows and doors would not be allowed. Within a certain distance from the property line, walls and these same openings would have to be rated as well. In addition, the Building Code has a height restriction but it is normally higher than zoning rules. Therefore, if zoning has increased requirements, they would supersede the Building Code.

Allowable Height and Size Based on the Building Code

The first limitation of the Building Code is building height. The International Residential Code (IRC) limits a house to a maximum of three stories. If you want a taller home, you must use the International Building Code (IBC). The IBC allows taller buildings (unlimited) but requires a much higher fire rating on the building materials and fire protection equipment such as sprinklers and fire resistant construction materials. The following excerpt from the IRC defines its scope by referring to the building height:

> *R101.2 Scope. The provisions of the International Residential Code for One- and Two-family Dwellings shall apply to the construction, alteration, movement, enlargement, replacement, repair, equipment, use and occupancy, location, removal and demolition of detached one- and two-family dwellings and townhouses not more than three stories above-grade in height with a separate means of egress and their accessory structures.*[1]

Required Setbacks

Normally, the exterior wall of a single-family home can be built at not less than 5 feet of the property line without any special considerations. You can even be at 3 feet from the property line and have a portion (25% maximum) of the wall space attributed to non-fire-rated windows and doors, but the exterior wall must have at least a 1-hour fire resistance rating. This affects a "green home" in that certain materials you might choose to use to build may not be rated for 1-hour fire resistive construction.

[1] *2006 International Residential Code, International Code Council Inc.*

Subdivision Conditions

In addition to zoning and building restrictions, there are sometimes rules within subdivisions or developments that further restrict building design. These are known by many terms, such as *restrictive covenants, subdivision restrictions,* or *community standards.* Common to a subdivision, these types of rules are normally established by the developer and made a condition of ownership by adding to a deed or by establishing in a community association's bylaws. Subdivision restrictions limit how your property may be used. This type of land use control is similar to a zoning control but often more restrictive. In addition, the restrictions may include controls that the local government is prohibited from enforcing. Subdivision restrictions can prevent owners from making changes to their lot that creates a look that conflicts with that of the remainder of the community. The intent of such a restriction is to preserve the continuity or general appearance of the neighborhood and preserve general property values. For example, a subdivision restriction might limit your choice of architecture or building style within the development. It might further limit the building's height or size. It may require or even limit the placement of a fence or material for the fence, restrict certain types of exterior building materials, and restrict use of antennas or satellite dishes. The purpose is the same as that of zoning: to preserve the look and feel of the community. It improves the long-term value of the investment but it also can affect your goal of having a green home. The materials you choose may not be allowed. The size of the home itself may be too small for the community. The use of energy-producing equipment such as solar panels or wind generators may be prohibited or restricted in some manner.

Review the restrictions ahead of time and evaluate if your green home is a good fit. Most subdivisions with restrictions like these have an architectural review committee composed of residents. If you are concerned, meet with them and go over your plans before you buy. It is better to know ahead of time than to find out after you have bought the land and have to change your building plans.

Endangered Species

Construction work could incidentally affect an endangered species. Under the authority of the Endangered Species Act, the U.S. Fish and Wildlife Service regulates the adverse impact that a construction project may have on endangered species. Certain wolf species, like the one shown in Figure 2-4, are examples of endangered species that need protection. A permit may be needed if your building project affects a threatened or endangered species. It is the builder's responsibility to determine if they are affected. The U.S. Fish and Wildlife Service and most state fish and game agencies can assist in determining if the area is affected. Birds are the most common protected species. The Chesapeake Bay Preservation Act (CBPA) is an example of environmental protection on both the environment and living things affected by pollution within such a protected area. The CBPA establishes building restrictions within certain distances from the edge of a protected area. Because of this, the buildable area of your lot may be further limited.

Conservation Easement or Wildlife Habitat

A conservation easement is a binding agreement between a landowner and either an organization or governmental jurisdiction (federal, state, or local) that restricts certain land uses in order to protect the environment from conditions resulting from that use. The easement restricts land use by limiting development and construction or farming or ranching activities that otherwise would occur in years to come. Because the ordinary use of the property is now limited, the market value of the property may change. Conservation easements protect the

Courtesy of iStock Photo

Figure 2-4 ■ Natural habitat for wildlife has dwindled because of construction. Certain areas are designated as wildlife preserves and protected from development.

Photo by Lynn Underwood

Figure 2-5 ■ A bird sanctuary would attract waterfowl and serve as a wildlife habitat.

natural beauty as well as the habitat for wildlife. If your parcel is particularly large, agreeing to an easement of this type would be a definitive statement of supporting the intent of *green* building. A bird sanctuary that attracts waterfowl like the one in Figure 2-5 is an example of a wildlife habitat. Future owners are also bound to the easement's legal agreements.

A simple solution for your new home is creating a backyard refuge for wildlife. A garden for natural wildlife would be another positive statement of living in harmony with fellow residents of this planet. It returns something to the animals that may have depended on the vegetation that grew where your house will be. But there is a benefit. You can watch wild birds and other animals visit.

Use of Natural Precipitation or Reuse of Gray Water

Water is essential to the lives of all living things. Most usable water comes to us ultimately from rainwater via the hydraulic cycle. Rainwater falls on your roof and surrounding hard surfaces quite regularly. Precipitation purifies water through evaporation and condensation except for some conditions of atmospheric pollution. In all cases, this rainwater is yours for the taking, although a few states like Colorado and Washington have laws affecting prior-owned water rights and prohibit water harvesting.

Legal Access

As a property owner, you will have use of your property with certain conditions. Do not assume that you will have legal access to property that you are considering. Just because you can drive up to or even walk onto the property does not mean that you may not be trespassing on private property to do so. A legal access is normally specified in the deed. A legal property survey should specify the access; however, that is not always the case. An easement is another method of providing this legal access. A word of caution: Easements can be permanent or temporary. Although an easement may exist, it may also cease to exist based on any number of conditions, including expiration of time. For example, you may request and be granted an easement from a neighbor to use a portion of his or her property for access while the primary access is blocked. In all likelihood that easement would have a time limit. If the easement was granted improperly or in error, it can be revoked. Title companies provide insurance policies to protect the legal interests of the buyer and the mortgage company. Legal access is a significant part of their review. However, even title searches may not bring some situations to light.

In any case, normally when you apply for a building permit, a determination is made whether or not the legal access is available. By then, it could be too late. Use legal assistance to determine if your access is legal and permanent before buying.

Legal access to the building site needs to be available (and convenient) for delivery of construction equipment or building material. Delivery vehicles and construction equipment will drive up to or even onto your lot. They will need legal access, so planning ahead is important. It will be important to know who maintains access, if it is reliable, and if a key is needed. You may encounter a locked gate on an access to your home site like the one shown in Figure 2-6. Consideration should also be given to how delivery of materials could comply with your plans to lightly touch the environment, if there is a possibility of trees being damaged to gain the access, and if there is a need to add any impermeable surfaces (concrete or asphalt) to allow vehicles to enter your property. Consider the negative impact on the environment that this type of surface would cause.

Utilities, Services, and Conveniences

Proximity of utilities to the site location is an important factor when planning to build. Cost and upheaval of natural soil are two considerations. A serious concern is the cost to install utility service to your new home. If the distance is significant, the cost may be exorbitant. In fact,

Courtesy of iStock Photo

Figure 2-6 ■ Legal access to your lot is not assumed unless clear in the deed or ownership records. An access road may be there when purchasing but if it is owned by another, your rights to use it may be withdrawn.

if a price on a great-looking piece of property is too low, this may be the reason. Check it out by calling the utility companies and asking for the cost of connection.

Development of Utility Infrastructure

Part of a green home is bringing as little disturbance as possible to the Earth. Consider how much excavation will be necessary to connect utilities to your home. Positioning your house on the lot in an optimum manner will minimize land disturbance. You can also plan the location of each utility access to a strategic point on your house closest to the meter or public utility delivery point. As an example, electricity is a very common utility that is essential to land development. Figure 2-7 illustrates an electrical service panel and meter base that serves that purpose.

Sanitary Sewer

Use of a community sanitary sewer system is a much more efficient means of eliminating natural waste than a private sewage disposal system (septic tank). It serves the greater community by providing sewage collection and treatment for all. Due to the cost of these systems, normally you

Courtesy of iStock Photo

Figure 2-7 ■ Utility services such as electricity or gas must be installed before or concurrent with the manufacture of the home.

are required to connect if service is available and near your lot to spread the cost across many subscribers. Check with the sewer purveyor regarding availability and cost before you select a lot.

Private Waste Sewer

If a public sanitary sewer is not available, you will need to provide a private sewage disposal system such as a septic tank with leach or drainage lines. This disturbs more of your natural soil and eliminates the use above ground where these are installed. For instance, you cannot build any structure over a septic tank or leach field. The disturbance of the soil is from excavating a large hole; burying an impermeable tank, sometimes concrete treated with fly ash; and digging one or more deep trenches (leach field) that stretch away from the tank. The leach lines are often filled with aggregate. A perforated pipe sits on top of the aggregate just below the surface. The septic system digests the contaminated sewerage in compartments of the tank, then discharges the (relatively) clean waste into the perforated pipe in the leach field.

Soil conditions for sewage disposal

In order to evaluate if the site can have a private sewerage system, a soil analysis of the drainage characteristics is necessary. The leach field allows the clean wastewater to percolate through the aggregate and into the soil on either side and be cleaned further through this action. In order for the size of the leach field to be established, the rate of percolation in the soil must be determined. This test is conducted by a qualified party such as a geotechnical engineer. The percolation rate and other factors will help the engineer design a system *if* it can be done on your property.

Storm Sewer

Storm water drains collect unusual accumulations of water from precipitation and carry the water away from your site. From an environmental perspective, the concern is on whether the construction activities will pollute the storm water runoff. There are measures to ameliorate the impact on pollution of this water. During construction, protect the inlet drains with filtration measures such as sandbags. If your lot has no storm water sewage system, the runoff must be accommodated, either by a channel away from the property or by on-site retention. Otherwise, it may contaminate the groundwater.

Domestic Water

Reports are available on water quality from the public water utility. If no report is available, have the water tested. Look for purity and content of parasites and bacteria. Basic water test kits evaluate content of bacteria, nitrates, pH, sodium, chloride, fluoride, sulfate, iron, manganese, total dissolved solids, and hardness.

Water treatment

Sometimes, a purification system is needed. A water treatment process generally involves removing unwelcome particles. Contaminants that are removed during the process of potable water treatment include bacteria, algae, viruses, fungi, minerals such as iron and sulfur, and artificial chemical pollutants.

■ SITE WORK AND LANDSCAPING CONSIDERATIONS

A green home is one that conserves energy but also has a minimal impact on the environment. By definition, when you build something, you disrupt the environment. It would be impossible to live in this world, providing the basic necessities, without affecting the environment. However, your green home can reflect a reduced impact. Nowhere will your impact be greater than in the direct effect on the Earth through excavation and site work. There are lots of considerations and trade-offs that will lead to the most appropriate *shade of green*.

Physical and Natural Conditions of the Property

Site condition has a significant effect on the ease of building a green home. Remember, every beneficial condition is one that will not have to be added during construction such as the need to cut and fill the terrain. Sometimes these conditions are not easily noticeable and may involve contamination or are affected by proximity to a waterway. The condition of the soil may be such that unwelcome soil admixtures are necessary to improve structural integrity. Prevailing natural conditions are important when alternative energy sources or passive heating methods are considered. Carefully check the site for all related conditions, including the following, for the effect on your plans.

Land Use History

Research to be sure that your site has never served as a waste or dumping ground. Consider the variety of products commonly used in homes—rural, suburban, or urban. These may be paint, fuel, oil, cleaning solvents, woodworking materials, batteries, and glues, as well as insecticide and weed killer. Control or the lack of control of these materials can become a big problem for us in modern society. Casual disposal of these materials threatens groundwater and air quality and contaminates soil. Sometimes portions of these products are lost through spillage, are buried or dumped, or are washed out onto a site. Control of this type of indiscriminate solid waste disposal on a home site improves long-term health for humans and the environment. Figure 2-8 illustrates the environmental impact of this indiscriminate waste disposal. Many people take the hazards created by this activity for granted. Improper use of these hazardous materials pollutes

Courtesy of photos.com

Figure 2-8 ■ Any building site could have been (or may still be) a dumping ground for waste and construction debris. This may cause damage depending on the chemical content of the material. Look for signs at your proposed site.

the land, water, and soil. Chemicals could also leak from containers and affect drinking water through groundwater contamination.

These examples of pollution could be obvious if trash or debris is above the surface. Detection may be difficult if the land was used as a landfill either with or without authorization. If garbage is present in any condition, removal could be expensive, especially if the material is a chemical, toxic, poison, or other hazardous waste. Pollution like this could affect groundwater, air, or even the soil itself. Consider this type of pollution as significant if it is toxic to you or the environment. Environmental pollution on a home site could range from household waste to automobile parts, agricultural residue, yard waste, construction waste, and by-products of illegal drug manufacturers.

Soil Debris

Inspect the entire surface of the lot for debris. Certain soil is not conducive to construction. Also, demolition of a building might be the reason that concrete is dumped in a convenient spot. Whatever the case, it would become your problem to remove the debris from the property. Trees or brush can conceal this type of activity over time, so look carefully.

Regulatory Control

It is necessary to know the regulatory control on the site you select. There could be restrictions, including a limitation of natural terrain or governmental protection or a naturally occurring element such as a mountain, ravine, embankment, ditch, farm field, elevation, depression, relief, or slope that could affect your plans. It is important to know if any adjoining land is owned by the federal, state, or local government or if any of the preceding entities have an interest in that property such as a plan to lay pipeline across it. A utility company could hold an easement across your property, preventing your use as well.

Waterway

There are numerous issues you face whenever a site is near or adjoins a waterway. The potential pollution to that waterway by a construction project is enormous. Additionally, the integrity of the body of water may be subject to change. You may purchase a beautiful waterfront lot on a small, private lake only to have that dam break and the lake disappear.

Wetlands

One part of living near a waterway is preservation of the natural wetlands. A natural wetland is where water is the main aspect that preserves the environment and its plant and animal life. These occur where the surface of the land connects to a body of water or where the land is covered by shallow water. Wetlands account for about 6% of the Earth's surface. Besides being among the world's most productive environments, they are a center of biological diversity where countless species of plants and animals live. Wetlands can occur at coasts, estuaries, flood planes, marshes, swamps, lakes, and even where rivers meet the sea. There are countless reasons to preserve wetlands, but some of these may inhibit your plans to build. For instance, there may be a setback from these wetlands that restricts where you can build on your land. Wetlands may be similar to the building site shown in Figure 2-9. In this case, construction activities may bring harm to a larger waterway. Also, your construction project may be required to take extraordinary measures to control your construction activities. A driveway may not be

Courtesy of photos.com

Figure 2-9 ■ Areas adjacent to waterways may be subject to wetland protection. Runoff or erosion from your construction site may adversely affect the waterway.

permitted where you desire or you may not be able to remove certain vegetation, even if it is decayed or harmful to your health.

Flood hazard zone

If your lot is within proximity to water or near the prevailing elevation of a waterway, it may be within a special flood hazard zone. These zones are established and maintained by the Federal Emergency Management Agency (FEMA) to mitigate damage to buildings caused by flooding. Most affected areas have been identified and are cataloged by different *zones* as mapped by the National Flood Insurance Program. These various zones relate to the expectation of flooding and thus the potential damage to buildings. According to FEMA, the flood hazard areas identified on the Flood Insurance Rate Map (FIRM) are identified as special flood hazard areas (SFHAs). SFHA is the area that will be flooded by an event having a 1% chance of being equaled or exceeded in any given year, thus, the 100-year storm event sometimes called the base-flood event. SFHAs are labeled as Zone A, Zone AO, Zone AH, Zones A1–A30, Zone AE, Zone A99, Zone AR, Zone AR/AE, Zone AR/AO, Zone AR/A1–A30, Zone AR/A, Zone V, Zone VE, and Zones V1–V30. Moderate flood hazard areas include Zone B or X and represent the 0.2% annual chance (or 500-year) flood.[2]

[2]*Courtesy of FEMA, www.fema.gov/*

Soil Report

Learn how to order and read soil reports. They are also known as geotechnical reports when related to construction activity. Geotechnical reports are normally prepared by a registered engineer experienced in soil science. This report may be required to obtain a building permit, depending on prevailing soil conditions. The report determines soil conditions that affect the existing soil's ability to support a building. These reports are normally required in areas with expansive or compressive soil. Normally, soil reports are required if the building foundation is to be supported by engineered fill material. Geotechnical reports may be required by the building official, perhaps before permits are. An engineer designs the foundation based on the results of a geotechnical report. The design (size, width, thickness, and depth) of the footing and the type of foundation system depend on this report unless satisfactory soil conditions prevail.

Prevailing Terrain

The usability of the prevailing terrain in its current condition to avoid using cut and fill practices is critical if you want to avoid introducing earth-moving equipment onto your building site. The more energy it takes to construct your new home, the less green it will be. Look for lots with prevailing conditions that match the desired finish grade appearance as much as possible.

One notable exception of the naturally flat terrain of a building lot is a *berm house.* This is an earth-sheltered house designed to be partially buried underground or have dirt piled up or *bermed* up against one or more of the exterior walls. This type of design could easily rely on the natural terrain that would otherwise appear to be a problem for a conventional house.

Other Natural Considerations

Use prevailing nature to your advantage. Instead of fighting nature, position your home to benefit from the natural site conditions that are in place and embrace nature. See how your house can fit within the prevailing conditions, rather than trying to alter natural conditions that have developed over thousands of years. The greatest hazards to any development occur when we do not recognize that natural events will continue to shape the landscape along the progression of time.

Sun

Track the solar path across your building site to see how your home could take advantage of this free natural resource that is totally renewable. Whether you plan a passive heat gain from south-facing windows, translucent panels, or photovoltaic cells for generation of electricity, you need to know if the lot conditions are optimum.

Wind

Wind is the irregular movement of air across the surface of the Earth. This driven air results from uneven heating of the Earth's surface. Air absorbs and carries energy from the heat drawn from the Earth. No matter what climate you are in, you will need to accommodate nature, exposure, and prevailing speed of wind at your lot. Of course, different climates have different approaches to meeting the conditions brought by wind. In a cold climate, wind can be a thief, robbing you of the energy you use to heat your home. However, it can be your friend, cooling your home on a hot day in an arid climate. Wind can also be harnessed to drive wind generators, developing electricity to use instead of traditional nonrenewable energy sources.

Wind power

While considering the use of wind generators, determine if there is any prohibition against tower installation. For example, there may be a height limitation or the need to stabilize with guy wires or bracing. Generators generally need to be high enough to gain a steady flow of wind. However, the generators are supported by towers and there are several concerns. Consider the wind generator in Figure 2-10. Excessive height on a wind generator may become the source of several concerns. First, zoning may limit the height. Second, being high, the same wind you want may cause your tower to be a hazard to you or your neighbor if it falls. General aviation rules may limit the placement of a tower as well.

Courtesy of photos.com

Figure 2-10 ■ Wind generators use natural forces to generate electricity. Although the payback is longer than some measures, as the cost of energy rises, the payback period will undoubtedly be shorter than first anticipated.

Adverse effects of wind

Wind can also have an adverse effect on home building. The land may be subject to erosion from heavy wind or dust storms. Determine if the lot provides any natural protection for the building site. Imagine the lot in the other three seasons and consider the effects of snow, rain, or hurricanes on your building.

Wind and heat

Prevailing wind can provide fresh air while avoiding loss of heat that is driven away from a house in wind wash during the heating season. A green building considers the surrounding climate and natural energy sources. In a cold area, vegetation (such as trees and brush) and surrounding terrain (hills and mountains), as well as other buildings, can help a building maintain its warmth, thus reducing energy consumption for heating. In hotter zones, the opposite may be the case. You may want air to flow through your home, allowing for natural ventilation. When you are lot shopping, look for the seasonal and daily path of the sun across the building site, the seasonal and daily wind-flow patterns, and the low areas where cold air can get trapped; and consider how the prevailing wind and sun patterns will affect your heating and cooling needs.

Green Landscaping

A green home has landscaping that matches the climate and regional conditions. If nature can help sustain your landscaping, then irrigation and excessive water use can be avoided.

Xeriscape

Xeriscape is landscaping that uses certain plants that can thrive with little supplemental irrigation. The term *xeriscape* is derived in part from the Greek word "xeros," which is Greek for "dry." Drought-tolerant plants include more than succulents and can enhance any landscaping plan. There are many examples of low water usage among plants. Low use grasses, turf alternatives, ground covers, hedges, bushes, shrubs, flowering plants, and shade trees are all drought tolerant. Xeriscape does not have to appear any different from normal landscaping. Figure 2-11 is an example of how this type of landscaping can blend in.

Tree Preservation

If trees or any other vegetation is to be saved during construction, the entire plant structure must be considered. In particular, a tree's root system sustains the tree. Damage to a tree's root system can be affected by more than direct excavation at or near the tree. Changes in surface grading, soil consolidation, and certainly construction in an area above the root system will cause distress and lead to poor growth or slow loss of the tree. There are many ways to protect the tree, including the simple act of marking and installing a barricade to prevent entry into the determined zone. This zone varies for each type of tree but usually is significantly greater than the drip zone (the area subject to drip from precipitation). Marking this zone will prevent anyone from digging arbitrarily or otherwise causing disturbance. This will prevent any driving or parking by construction vehicles as well.

However, if digging for pipe or conduit installation is necessary, Figure 2-12 illustrates how this can be accomplished while still maintaining the health of the tree. Protect the root system from damage by digging or boring away from the major root system.

Courtesy of iStock Photo

Figure 2-11 ■ Xeriscape tends to use plants that are native and thus relies on prevailing rainfall for growth. Stone or gravel added to the landscaping provides accent where needed.

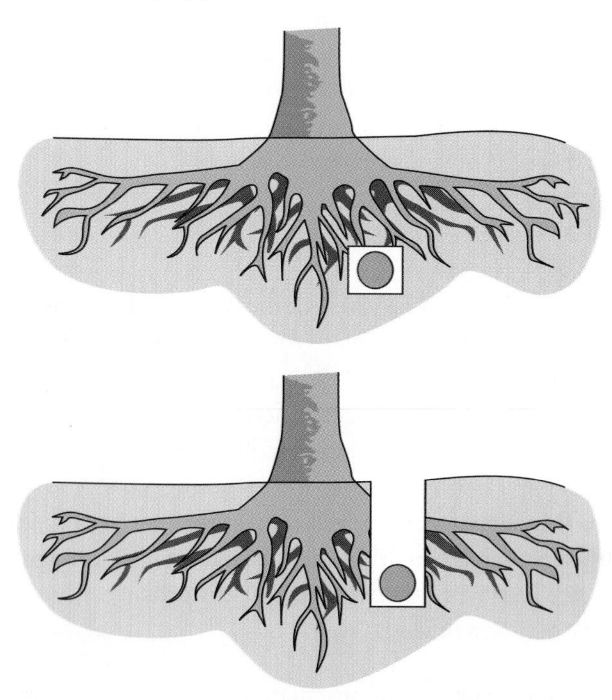

Courtesy of Southface/EarthCraft House

Figure 2-12 ■ Installing pipe or other building materials underground can have a detrimental effect on trees if roots are severed or damaged. Careful digging around trees will preserve the integrity of the root system.

Tree removal

Inevitably, a tree may need to be removed during construction. Finding a responsible way to remove and dispose of the tree is important. Urban or suburban building sites are not the normal locations to find lumber to process. There are many problems associated besides the quantity levels available at a single spot. Accidental damage of utility lines such as water, sewage, electrical, telephone, or gas would be expensive to repair and may even involve stiff fines. There is always the potential for embedded objects such as metal fragments that could damage a chain saw. Removal of a tree can be difficult with other residential homes within falling distance. Further, the hauling of a few trees to a sawmill seems less than cost-effective. However, portable sawmills make it worthwhile to come to a site such as a lot to be developed for the value of just a few trees, especially if they are sought after by woodworkers. At the very minimum, you can reuse some of the lumber in landscaping or on-site fencing. Be resourceful and efficient. You might even use some lumber in your construction project if permitted by the local building department, which may require evidence of the structural qualities of the lumber.

Lumber

Lumber must be evaluated and accepted before it can be considered usable in a building project. First, a freshly cut tree is filled with water. The moisture content must be reduced to 19% or less before it is acceptable for use in construction. You can allow it to air-dry after milling, but its moisture content must be established. Drying wood naturally may take a year or more. Additionally, the structural integrity of the wood must be evaluated. At lumber mills, it is generally done in one of two ways: visual grading or machine stress rating. The machine stress rating is more accurate but costly. Visual rating is acceptable. However, you must find someone qualified with sufficient experience to grade lumber who can satisfy the local building official. A structural engineer with the proper credentials may be able to provide this service.

Erosion and Sediment Control

You may consider a plan to care for the surface and subsurface soil around your construction project. An erosion and sediment control plan includes steps to avoid environmental damage adjacent to your construction project. It seeks to limit the environmental harm during construction and is essential in some locations such as those near or adjacent to waterways or near environmentally sensitive environments. In other areas, it is good practice to protect the neighborhood from the detrimental effects of your construction activities. Begin by thinking it through. First, you will bring equipment to clear land of vegetation and build a house. Think about the impact that will have on the environment as a whole. Consider where the rain will go that normally percolates into the soil where your house will be. It is possible that the exposed soil will erode and wash away. It could flow to the neighbor's yard or the street. Plan ahead so as to prevent any unnecessary damage.

There are three basic things you can do to limit erosion damage caused by construction:

1. Erect a silt (siltation) fence around the perimeter of the lot or building site. This is a low-profile barrier that prevents soil from discharging away from the site such as depicted in Figure 2-13.
2. Install protection at all storm water drainage inlets that filters out sediment that would normally drain into the waste disposal system. This can be something as simple as a sandbag filled with filtration material covering the inlet.

Courtesy of iStock Photo

Figure 2-13 ■ Fabric erected around the job site prevents erosion, controlling the movement of topsoil, and prevents runoff from discharging harmful materials in waterways. This is required in some areas during construction.

3. Establish a practice of cleaning soil off vehicle tires *before* they leave your site. It may not seem like much, but erosion such as this causes serious long-term problems. Also, laws in some states require just such practices to limit erosion.

Soil Care

For proper care of your soil, work out a landscape and soil conservation plan. Work to restore or even improve the natural conditions that existed on your land before you started your project. You can do this in several ways, including establishing a conservation area around your property. Particularly in suburban and rural areas, you can also encourage the wildlife to return by building an inviting environment for them.

Rainfall

Storm water and rainwater harvesting is a significant part of a green home. If you can collect and reuse water, a savings is found in two ways: (1) you save the natural resource itself, and (2) you save the energy cost to deliver this resource to your home. Even if you do not choose to attempt collection of potable water, the least you can do is to save it for irrigation purposes. Rainwater harvesting reduces the demand on water supply and concurrently reduces the potential for flooding downstream and the potential loss through erosion of any topsoil in the path of the runoff. Rain harvesting is a burgeoning business. Collection devices range from simple barrels to large, underground containers with pumps. However, be aware of local and state laws that restrict the harvesting of water if water rights downstream of the collection point are reserved or protected.

Gray water irrigation

Gray water from a home is any used water from a source other than a toilet or urinal. It includes kitchen and laundry sinks, washing machines, bathtubs, showers, lavatories, and even outdoor

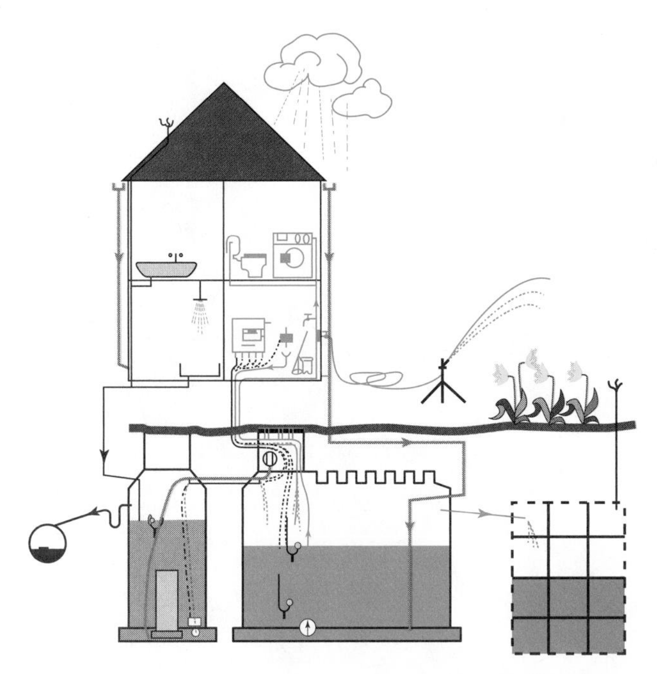

Delmar/Cengage Learning

Figure 2-14 ■ Gray water systems make use of lightly used potable water within a home. Instead of going down the drain, water is reused in some manner. This system uses both gray water system combined with water harvesting system for reuse together.

hose bibs. Reuse of this type of water is mainstream and even approved by the Plumbing Code under certain conditions.[3] Reuse of this kind decreases strain on sewage facilities with less effluent being delivered for cleaning. In any case, a gray water system requires planning. Figure 2-14 illustrates a rough sketch that would occur early in the planning stage for a gray water system. This one has made use of both a gray water system and rain harvesting simultaneously.

Alternatives to Hardscape

Concrete or asphalt paving is hardscape. It does not allow water to permeate into the ground. This type of nonpermeable surface increases runoff from rain, causing flooding and erosion of top soil. It also causes increased pollution caused by the runoff. If rain and other precipitation can permeate into the soil, natural conditions will prevail. Any attempt to provide an alternative to concrete or asphalt paving represents environmental design. There are many alternatives to traditional hard surfaces.

Permeable paving

Permeable or porous paving materials can be a beneficial alternative to the traditional hard surfaces of poured concrete, concrete pavers, bituminous asphalt, and similar materials used in driveways, walkways, sidewalks, and even patios. There are several types of permeable paving materials: porous pavement, gravel, grass pavement, spaced pavers on sand, brick, grid or lattice rigid plastic, crushed stone, bark chips, mulch, or cobbles. There are numerous innovative materials that provide the traditional hard surface and appearance while being fully permeable. Resin pavement over permeable asphalt is just one modern alternative material that allows the hard surface needed for vehicular traffic while allowing for water infiltration. These permeable paving alternatives do not have to appear unsightly. Figure 2-15 illustrates a permeable paving option that simulates cobblestone.

[3]*2006 International Plumbing Code, International Code Council Inc.*

Courtesy of photos.com

Figure 2-15 ■ One solution to preventing excess runoff from too much impermeable paving material is to use porous-type paving. Even though a hard surface is needed for parking, permeable paving can be a substitute, allowing water to infiltrate into the soil and water table instead of carrying soil away as it drains into a street.

Some newer concrete mixes are fully permeable, allowing all water to pass through. However, it is necessary to prepare the sub-base with a gravel layer to allow for the drainage to penetrate the water table.

Paving alternatives

Alternative paving design is an easy way to reduce the impact caused by hard surfaces. It can involve driveway, walkway, or sidewalk redesign. There are many alternative designs that improve effects of hard surfaces. In a residential community, shared parking spaces or shared driveways reduce the need for so many expansive hard surfaces. Eliminating walkways in favor of stepping stones will help reduce this effect. If you avoid a sidewalk installation on both sides of the street, you reduce by more than half the influence of a traditional hard surface. You could have on-street parking, if permitted by the jurisdiction (some zoning requires off-street parking). At the very least, you could design any impervious hard surface so as to drain toward vegetation such as lawns, a flower garden, or shrubbery that has a permeable surface.

■ BEFORE YOU DECIDE . . . REFLECTIONS AND CONSIDERATIONS

✔ Site selection is the first and, in some ways, the most critical step in building a green home.

- Proper site selection allows for the best possible use of all natural resources and ensures that any development is built in harmony with the natural environment.

- Appropriate site selection creates the opportunity for optimum design according to climate and careful development with respect to environmental conditions.
- With many considerations, the determination of the site you select will likely be a decision based on numerous compromises because it will be a home for more than one.
- Proximity to available services is critical to avoid long-term vehicular use of fossil fuels.

✔ A guiding principle should be consideration for leaving the natural conditions as they are or to improve the site environment with the natural look.

✔ The location, orientation, and landscaping of a home building project all affect the local population, environment, regional transportation, and energy use.

✔ Rules such as zoning, Building Code, neighborhood restrictions, legal access, and endangered species must be observed.

✔ Utility services such as electricity, gas, water, sewage, and communications must be considered.

✔ Physical condition of the lot affects the development. Factors include soil type and bearing value, water table and wetland, flood hazard potential, adequate prevailing terrain, and appropriateness for your home design.

✔ Climate and other natural considerations such as wind patterns, rainfall, solar orientation and access, frost potential, temperature, tree preservation, and natural vegetation are important aspects of green building planning and design.

✔ Erosion control, soil care (and necessary soil treatment), landscaping, permeable paving, and the potential for gray water usage are essential to consider when planning a green home design.

FOR MORE INFORMATION

Environmental Protection Agency (EPA)
http://www.epa.gov/

Federal Emergency Management Agency (FEMA)
http://www.fema.gov/

International Code Council
http://www.iccsafe.org/

The National Wildlife Federation
http://www.nwf.org/

U.S. Fish and Wildlife Services
http://www.fws.gov/

Chapter 3

GREEN HOMES, SUSTAINABLE DESIGN, AND THE BUILDING CODE

SAFETY CONSIDERATIONS

Even though a major concern when building green is the environment, the primary reason for building a home is personal shelter from the natural environment as well as from man-made conditions. Central to this theme is safety. The building has to stand up to natural elements that are likely to exist during its lifetime. Things like wind, hurricanes, earthquakes, solar heat, cold weather, rain, flooding, and wind-borne debris all adversely affect a building. Keep the concerns for the environment in mind, but, first and foremost, a home should be built safe and strong enough to last.

The Building Code

When constructing a new house, you will be required to comply with provisions of the Building Code. The International Residential Code (IRC) is the most likely code book you will use. This model code book is most likely adopted in your jurisdiction, perhaps with some local or regional amendments that reflect local customs for construction. The IRC contains provisions for architecture, structure, energy efficiency, plumbing, mechanics, and fuel gas and electric systems for homes up to three stories in height (for homes taller than that, the International Building Code [IBC] must be used). Figure 3-1 shows the IRC, along with several other International Codes. It is a standalone document with provisions for building, plumbing, electrical, fuel gas, and mechanical installations. The IRC is filled with prescriptive provisions and is like a *cookbook* with graphical instruction on proper installation of a myriad of materials, components, equipment, and fixtures. The following are a few of the aspects of construction regulated by the IRC. Although building a *green home* is your goal, the inspector will regard it as a construction of a safe building first and a green home second. You must meet these minimum requirements to satisfy the safety conditions before you seek to reach any higher standard.

The following checklists include some of the most common aspects of the code that must be satisfied before a permit is issued or before an inspection is passed. This is not an exhaustive list. These requirements are the safeguard by which dangerous conditions are avoided. However, they are prescriptive in

Delmar/Cengage Learning

Figure 3-1 ■ Recent published editions of the International Codes (I-Codes).

nature. They regulate the most ordinary building materials and the most conventional of construction methods. These prescriptive regulations relate to techniques using materials such as conventional wood frame floors, wall and roof framing, or concrete/masonry wall assembly in the construction process. Generally, the IRC limits its analysis to building materials, including wood framing, light gauge steel framing, masonry, and concrete (including insulated concrete forms).

Architectural Requirements

Chapter Three of the *International Residential Code for One- and Two-Family Dwellings*[1] sets out many safety requirements based on architectural design. First, a building's height is limited by the Code to three stories. Then the ability to safely exit the building is specified for normal conditions as well as in an emergency. The following are representative of more common requirements:

- Natural lighting for each room should be at least 8% of the floor area.
- Natural ventilation should be at least 4% of the floor area for operable portions of the windows.

[1]*2006 International Residential Code, International Code Council Inc.*

- The minimum dimension should be at least 7 feet in any horizontal length for any habitable room.
- Ceiling height should be at least 7 feet in habitable rooms.
- Smoke detectors should be on each floor level, in each bedroom, and in each hallway leading to a bedroom.
- Door and window sizes should be listed and labeled for Energy Code compliance.
- Each home must be equipped with facilities capable of heating rooms to 68°F.
- Attic access must be 22 × 30 inches where the roof is at least 30 inches above the ceiling or where equipment is in the attic.
- Attached garages must be separated from the dwelling by ½ inch of drywall and solid core or a rated door.
- Emergency egress must be provided from bedrooms with 5.0 square feet (5.7 square feet above the grade level) operable portion and show a maximum sill height of 44 inches above the finish floor.
- Shower enclosures must have a minimum dimension of 30 inches and a floor area of 900 square inches.
- All glass bathtub enclosures shall be safety glazing. This includes windows within the shower below 60 inches above the floor.
- Each water closet must have a clear space not less than 30 inches wide and not less than 24 inches in front.
- For staircases, rise of the steps should be no more than 7¾ inches; run of tread should be a minimum of 10 inches.
- Required exit doorways must be 3 feet wide by 6 feet 8 inches tall.[2]

Structural Requirements

Structural considerations are essential to code compliance. Structural integrity deals with the foundation, walls, floors, roof, and all supporting elements. This includes two forces on buildings: gravity loads that are vertical and lateral forces (wind and seismic) that are generally horizontal. Gravity loads are overcome with proper footings and foundations and wall frames. Lateral forces are overcome with lateral bracing. This bracing can be imagined as if gravity acted sideways. Wind forces rack buildings along the lines parallel to their line of force (laterally) and are resisted by lateral bracing that resolves these forces into footings and foundations. The bracing can be as simple as plywood siding on a stud wall frame with anchoring systems.

Structural foundation

The foundation supports the weight of the entire structure. The footing width and depth are essential for the weight of the structure. As seen in Figure 3-2, sometimes reinforcing steel is installed to accommodate future masonry foundation walls. Numerous requirements must be completed to pass inspection. The following standards, set forth by the IRC for construction and composition, are just an example of the Code requirements reviewed by an inspector:

[2] *2006 International Residential Code, International Code Council, Inc.*

Courtesy of iStock Photo

Figure 3-2 ■ Footing must be the proper width and depth, free of debris and loose soil. Any required reinforcing must be installed properly.

- Soil must be protected against termite damage.
- Foundations must be capable of transmitting all loads to the supporting soil.
- In areas likely to have expansive, compressible, shifting, or similar soils, a soil report may be required.
- Concrete exposed and subject to weathering must be air entrained.
- All exterior walls must be supported on continuous solid concrete footings.
- Braced walls required for lateral forces must be supported by continuous reinforced footings.
- Foundations with stem walls and slab-on-grade must be reinforced with steel bars.
- Exterior footings and foundation systems must extend below the frost line or 12 inches below soil.
- Shallow footings must be protected from frost with insulation.
- The top of footings must be level and the bottom of footings must not have a slope exceeding 1:10.
- A wood frame wall must be anchored to the foundation with ½-inch anchor bolts at 6 feet on center (O.C.).
- Anchor bolts must be embedded 7 inches into concrete or masonry foundation.
- Footings adjacent to slopes must meet the setback requirements (3:1).[3]

[3]*2006 International Residential Code, International Code Council Inc.*

Courtesy of iStock Photo

Figure 3-3 ■ Installed wood floor joists must be properly spaced, blocked to resist rotation, and connected to a supporting wall to transfer loads.

Structural floor framing

Before you call for a floor framing inspection, there are several things you should review or verify to improve your odds of passing an inspection. The inspector will be looking for several critical items with a focus on structural integrity. The floor framing installation in Figure 3-3 illustrates a typical job site ready for inspection. Notice first that the job is clean. Then notice that all the floor joists are installed on the proper center spacing and have the appropriate lap and blocking. The following is an abbreviated list of what an inspector will be verifying:

- The species of wood being used must meet strength properties and grade.
- Bearing walls must match foundation plans for support by footings.
- Repetitive members such as floor joists must meet the minimum design for span and center spacing.
- Headers in wall frames must meet the minimum design for span and bearing capacity.
- Wood floor joists bearing on wood must have at least 1½-inch bearing.

Courtesy of Miller Custom Homes

Figure 3-4 ■ Proper roof framing installation will depend on slope and method of framing. In ridge board framing, rafters oppose each other at the ridge board and are connected with rafter and collar ties and normally a ceiling joist.

Structural roof framing

As with other aspects of framing, the roof framing must meet certain standards of construction. When the roof framing is installed, an additional inspection of the structural integrity of that system is required. Notice in Figure 3-4 that rafters are installed with proper center spacing and that roof sheathing is properly connected to the rafters. Many inspection items are verified at this time. The following is a guide to the critical aspects of that inspection:

- The species of wood being used must meet strength properties and grade.
- Bearing walls must match the wall framing plans for support of the roof member design for span and center spacing.
- Wood rafters bearing on wood posts or a wall frame must have at least 1½-inch bearing.
- Bearing walls must be in support of roof framing members.
- Rafters must be prevented from rotating with blocking at points of support.
- Rafters must be within span limitations.
- Girders and beams must be within span limitations.

Structural wall framing

Just like floor or roof framing, wall framing must be inspected. Notice in Figure 3-5 that the studs are at the proper center spacing in the wall frame and that electrical wire is

Courtesy of Miller Custom Homes

Figure 3-5 ■ Elements of a wood-framed wall. Required headers will be supported by trimmer or jack studs. A double top plate will be installed unless a single plate has the required connectors.

installed in the wall and ready for inspection. The following are a few items reviewed by the inspector:

- The species of wood being used must meet strength properties and grade.
- Bearing walls must match the wall framing plans for support of the roof member design for span and center spacing.
- Bearing walls must be in support of floor or roof framing members.
- Girders and beams must be within span limitations.
- Wall bracing must be provided for exterior bearing and nonbearing walls.
- Shear wall panels must be provided at every corner and every 25-foot interval.
- Required fire blocking for wall frames must be installed.

Energy Efficiency

Your green home must meet (or even exceed) basic Energy Code requirements. Figure 3-6 illustrates an insulated wall with a vapor barrier. There are other basic inspection items that must be installed in order to pass this inspection:

- Door and window sizes should be labeled by the National Fenestration Rating Council for Energy Code compliance.
- Walls, crawl spaces, floors, and attics must have proper insulation according to the zone.

Photo by Lynn Underwood

Figure 3-6 ■ The thermal performance of the energy-efficient green home will have more than a minimum amount of insulation. It will be installed in a professional manner and will have a vapor barrier installed in an approved manner.

- Heating equipment should be sized properly.
- The foundation or slab should be properly insulated.
- Vapor barrier should be installed on the warm (in winter) side of all insulation.
- Cross ventilation should be provided in the attic.
- Heating and cooling ductwork needs to be insulated properly.

Mechanical Requirements

Mechanical and fuel-fired equipment must be tested and inspected prior to being concealed with drywall. A fuel-fired appliance such as heating, ventilation, and air-conditioning (HVAC) equipment, exhaust ducts, vents, or even a fireplace, as illustrated in Figure 3-7, must be inspected for fuel supply, exhaust vent, combustion air, and the following items:

- Appliances installed in garages shall be mounted on platforms at least 18 inches above the floor and per the listing.

Photo by Lynn Underwood

Figure 3-7 ■ Mechanical equipment that includes gas- or fuel-fed fireplaces must be inspected and approved prior to use.

- Heating or cooling equipment must be properly sized according to recognized standards.
- Heat pump and other mechanical equipment must be labeled.
- Appliance installation methods must comply with the manufacturer's listing.
- Appliances in the garage or carport must be protected from damage due to impact.
- Access, clearances, and working space are required around HVAC equipment.
- Condensation drains must be installed for HVAC units.
- Exhaust fans for bathrooms, water closet compartments, and laundry rooms must be installed.
- Access is required if heating or air-conditioning equipment is in the attic.

Courtesy of photos.com

Figure 3-8 ■ Gas pipe must be checked for tightness and its ability to hold pressure. Gas equipment must be checked for proper support, access, clearance, and exhaust vents.

- If heating or air-conditioning equipment is in the attic, provide a catwalk and working platform.
- If heating or air-conditioning equipment is in the attic, show a switched light and service outlet.

Gas Equipment

Pipe that carries fuel such as natural gas or liquefied propane must be tested for tightness to avoid leaks of a dangerous gas. The piping in Figure 3-8 illustrates a complex gas pipe assembly. Check to be sure that the combustion air meets the needed requirements for a fuel-fired appliance; that both combustion and relief air are present for gas equipment; and that chimneys, vents, and vent connectors are installed where required.

Plumbing Requirements

Water and drain, waste, and vent (DWV) pipe must be installed correctly to avoid any unsanitary conditions. Figure 3-9 illustrates a typical DWV installation with a pipe and fittings that provide for the sanitary disposal of waste. The following is a brief list of common inspection items that may be reviewed during a rough-in plumbing inspection:

- Water service pipe size material is based on water pressure, demand, and distance to the water meter.
- Minimum water supply pressure must be at least 40 pounds per square inch (PSI).
- Maximum water supply pressure is 80 PSI without a pressure reducer.

Courtesy of photos.com

Figure 3-9 ■ Drain, waste, and vent pipe must maintain a watertight test. In addition, the water supply pipe must be able to maintain a test of the prevailing water pressure.

- A pan is required under the water heater with a discharge pipe.
- A pressure and temperature relief valve and discharge line are required at the water heater.
- Showers and tub-shower combinations must be provided a thermostatic mixing valve.
- Hose bibs must be located on the plan and noted to have backflow prevention.
- Specify that the waste line be as high as possible under the countertop for dishwashers.
- Specify that the shower be at least 30 inches in any dimension and provide shutoff valves for all fixtures.
- Show the size of the trap for each fixture and demonstrate compliance.
- Specify that the distance between traps and vents is proper.

Electrical Requirements

An important part of a green home is the safety related to electrical equipment and wiring. Poor installation could result in a fire or electrocution. The electrical power disconnect and service panel are the hub for this installation. Figure 3-10 illustrates a typical service panel with circuit breakers. The following is a checklist of items inspected during electrical installation:

- At least one switched light or outlet must be provided in bathrooms, hallways, and stairways.
- At least one switch-controlled light or outlet must be provided in all habitable rooms.

Courtesy of iStock Photo

Figure 3-10 ■ Electrical service panel must have proper wiring connections and be grounded. Wire must be installed properly and protected from damage with nail plates within the framing member.

- At least one switch-controlled light must be provided in attached garages and outdoor entrances.
- Every 12 feet of wall space within a room must be provided with an outlet.
- At least one receptacle outlet must be provided in hallways 10 or more feet in length.
- An exterior weatherproof GFCI outlet must be provided at grade level at both the front and the back of house.
- A GFCI receptacle outlets shall be provided at each kitchen counter space wider than 12 inches.
- GFCI receptacle outlet shall be provided adjacent to the basin in each bathroom.
- Bathroom outlets shall be on a separate 20 ampere circuit with no other outlets.
- Receptacles in bathrooms, kitchens, and unfinished basements must be GFCI or arc fault types.
- Receptacles in garages or carports must be GFCI types.
- Two-appliance circuits must be provided to serve the kitchen, breakfast, and dining room.

- Such circuits shall have no other outlets.
- A convenience outlet and light switch are required for attic heating and air-conditioning equipment.
- A service outlet for HVAC equipment must be provided within 25 feet of equipment.
- Where ceiling fans are to be installed, an approved outlet box is required.

ALTERNATIVE MATERIALS AND METHODS OF CONSTRUCTION

The prescriptive specifications refer to conventional materials accepted by the Code. However, there is material that is not regulated in the IRC. You can still use this Code to build a *green home* out of sustainable materials that are not within the Code. The building official is granted the authority to accept an alternative type of material even if it is not within the traditional list of materials regulated by the Code.

> *104.11 Alternative materials, design and methods of construction and equipment. The provisions of this code are not intended to prevent the installation of any material or to prohibit any design or method of construction not specifically prescribed by this code, provided that any such alternative has been approved. An alternative material, design or method of construction shall be approved where the building official finds that the proposed design is satisfactory and complies with the intent of the provisions of this code, and that the material, method or work offered is, for the purpose intended, at least the equivalent of that prescribed in this code. Compliance with the specific performance-based provisions of the International Codes in lieu of specific requirements of this code shall also be permitted as an alternate.*[4]

This language has been in the model codes for almost a century. The purpose is to allow the building official the authority to accept new materials that may be invented in the intervening time between editions of the Code. Codes almost always lag behind modern technology and innovative advancements. These aspects of the Code have been preserved to allow you to build with a material that suits your intent—to build *green*. Because safety is paramount in the Codes, the building official should verify that the design is satisfactory and complies with the Code through testing.

> *R104.11.1 Tests. Whenever there is insufficient evidence of compliance with the provisions of this code, or evidence that a material or method does not conform to the requirements of this code, or in order to substantiate claims for alternative materials or methods, the building official shall have the authority to require tests as evidence of compliance to be made at no expense to the jurisdiction. Test methods shall be as specified in this code or by other recognized test standards. In the absence of recognized and accepted test methods, the building official shall approve the testing procedures. Tests shall be performed by an approved agency. Reports of such tests shall be retained by the building official for the period required for retention of public records.*[5]

[4] *2006 International Residential Code, International Code Council, Inc., Section R104.11*
[5] *2006 International Building Code, International Code Council, Inc., Section R104.11.1, R602.3.2 and Section 1604.1*

It will be up to you, the developer or owner, to prove to the building official that your proposed material or manner of building is safe. You can do that any one of several ways, as long as they are acceptable to the building official. The most common practice is to have a registered design professional (architect or engineer) perform an analysis and substantiate the design in the form of a drawn plan that may include structural calculations. Engineering and architecture are included in the science of construction. The elements of structural design are much more elaborate than the prescriptive aspects of the IRC. In addition, there may be an independent evaluation based on testing that is already available. This testing may have an engineering analysis as part of the evaluation. That testing may satisfy the conditions referenced in the Code.

Another avenue allowed by the Code is the use of performance-based provisions. There are two ways that Code language appears: prescriptive and performance based. Prescriptive language, as is written in Section R602.3.2 of the IRC, is very tightly defined with no real deviation from those parameters.

> *Wood stud walls shall be capped with a double top plate installed to provide overlapping at corners and intersections with bearing partitions. End joints in top plates shall be offset at least 24 inches (610 mm). Joints in plates need not occur over studs. Plates shall be not less than 2-inches (51 mm) nominal thickness and have a width at least equal to the width of the studs.*[6]

A performance-based provision as stated in Section 1604 of the IBC has a more broad definition, however.

> *Buildings and other structures, and parts thereof, shall be designed and constructed to support safely the factored loads in load combinations defined in this code without exceeding the appropriate strength limit states for the materials of construction.*
> *Alternatively, buildings and other structures, and parts thereof, shall be designed and constructed to support safely the nominal loads in load combinations defined in this code without exceeding the appropriate specified allowable stresses for the materials of construction.*[7]

The aspect of a performance code is that you have another option—that of satisfying the intent of the Code, which is safety. Do not be discouraged by your first denial for a permit. This provision in the Code allows for the use of any *suitable* material.

■ MODIFICATIONS TO THE BUILDING CODE

There is another provision of the Code that you may find useful. The term *modification* means "to alter, change, amend, adjust or adapt." In the Building Code, it means to alter, change, amend, adjust, or adapt the Code itself. It allows the building official to *modify* any provision of the Code whenever there are *practical difficulties* in applying the Code.

> *Wherever there are practical difficulties involved in carrying out the provisions of this code, the building official shall have the authority to grant modifications for individual*

[6]2006 International Building Code, International Code Council, Inc., Section R104.11.1, R602.3.2 and Section 1604.1
[7]Ibid

cases, provided the building official shall first find that special individual reason makes the strict letter of this code impractical and the modification is in compliance with the intent and purpose of this code and that such modification does not lessen health, life and fire safety requirements or structural. The details of action granting modifications shall be recorded and entered in the files of the department of building safety.[8]

This provision has the effect of allowing the building official to alter *any* requirement in the Code but for a specific set of reasons and with the charge that no decrease in safety should arise from the action. This is an additional tool that allows you some liberty to use alternative means when any prescriptive (or performance) Code seems to block your way. This language has been carefully crafted to allow a freedom of design for possibilities that a Code could not possibly predict yet still maintain the stipulation that the basic tenant of the Code be met.

EXPLORING THE POSSIBILITIES: MANUFACTURED ALTERNATIVE MATERIALS

The Building Code is in a constant state of catch up with new technologies. Manufacturers, through research and development, make a new product or building material and promote it at home shows. The contractor or owner-builder tests it and it eventually is considered to be mainstream. We adopt it in the Code and establish conditions for its use. That sequence takes years if not decades.

Material Approvals

An example of a newly invented building material is the insulated concrete form (ICF) wall systems. These are forms that are used to create a concrete wall. The form itself is made of insulated polystyrene blocks that interlock to create an in-place wall form. The forms are left in place after pouring the concrete to create an insulating barrier on both sides. The result is a reinforced concrete wall with insulation on both the inside and outside. ICF wall systems had been around in one form or another for 25 years before the Building Code regulated this material. In the year 2000, the IRC placed it in the Code, allowing it to be prescriptive in one of three designs.

International Accreditation Service and International Evaluation Service

There are other options for accepting new materials that are built into the Building Code system. The Code allows testing as a means of demonstrating that your building material meets safety requirements. The only way manufactured products can enter the mainstream of construction materials is through approvals by testing and evaluation. Testing and evaluation are normally performed by an independent testing laboratory. The tests are standardized, meeting the conditions for other types of materials. Although these tests are normally paid for by the manufacturer,

[8]*2006 International Residential Code, International Code Council, Inc., Section R104.10*

the evaluation is not a rubber stamp. The laboratory is independent and normally accredited by the International Code Council (ICC) or its accreditation service affiliate, the International Accreditation Service (IAS). The IAS is a nonprofit, public-benefit corporation and a leading accreditation body in the United States. Internationally recognized, accreditation by the IAS makes it easier to market products and services both inside and outside the United States.

Product manufacturers are facing increased demand for impartial verification that their products are certified to be safe and to meet industry standards. If the products are to be readily accepted, the verification must be by certification agencies that are trusted and widely recognized. IAS accreditation provides the credentials that ensure that an agency's certification marks will have full acceptance in the marketplace and by governmental agencies that regulate product acceptance.

The IAS assesses and accredits competent product certification agencies, testing and calibration laboratories, special inspection agencies, building departments, inspection agencies, field evaluation bodies, and fabricator inspection programs. In operation since 1975, the IAS is one of the oldest accreditation bodies of its type and is a nonprofit, public-benefit corporation recognized worldwide.

The International Evaluation Service (ICC-ES) is intended to allow for the innovation of new and better products that improve performance or reduce the cost of construction. When a new product is developed, it must be tested by one of these accredited laboratories. When this new product is listed as approved by the ICC-ES, it is essentially a part of the Code.

In addition, the ICC-ES is the leader in the United States in evaluating building products for compliance with the Code. The ICC-ES does technical evaluations of building products, components, methods, and materials. The evaluation process culminates with the issuance of reports on Code compliance, which are made available free of charge to Code officials, contractors, specifiers, architects, engineers, and anyone else with an interest in the building industry and construction. Evaluation reports provided by the ICC-ES are evidence that products and systems meet Code requirements.

The International Code Council's Evaluation Service has created a new program that seeks to verify the attributes of a product or material that touts sustainability or green attributes. The Sustainable Attributes Verification Evaluation or SAVE seeks to provide reliable information about claims made by manufacturers of a product's *sustainability* or *greenness*. This service also allows manufacturers the opportunity to have their products evaluated and verified by an independent third party. The evaluation analyzes these claims as well as manufacturing processes and testing. The final listing is based on carefully developed guidelines that support genuine sustainability. The Web site where this is discussed is: **http://saveprogram.icc-es.org/**

Project Start-up

Your project should start by considering what materials to use. You will likely see many materials, some traditional and some alternative. If you select a new alternative material, ask the manufacturer about the product testing and evaluation before you give it serious thought. Otherwise, you will likely face these same questions when obtaining a permit. Without those testing data and approvals, it will be difficult and expensive to prove that the innovative building material is safe for your project. You will likely need guidance from a registered design professional and a testing laboratory. The cost of performing your own product evaluation could be significant, often much more than the cost of the materials themselves. Rely on a manufacturer whose product has already been listed and tested and has the approval of an accredited testing laboratory. That approval will save you time, aggravation, and money. It is the correct process to demonstrate product safety for modern technologies without a proven track record.

NONMANUFACTURED ALTERNATIVE BUILDING MATERIALS

There are other building materials considered as *alternative materials* for building construction that are not referenced in the Code. Many of these materials are being used every day and accepted by a building inspector because they have been traditionally used for centuries in that region. For instance, adobe construction is used all over the southwestern United States. At least two states have adopted an Adobe Code to set standards and manifest adobe's worthiness as a material. However, there are other cases as well. Many of these materials are used in sustainable design because of the low impact on the environment.

Evaluation and Acceptance

Products such as adobe, rammed earth, straw bale, bamboo, and cob construction and the use of techniques like pozzolans (mixture of lime and fine volcanic ash), sandbags, straw clay blocks, earthen plasters, and even paper blocks are considered marginal by building inspectors mainly because they have not seen them perform as often as they have seen a wood stud wall frame used. There are concerns with all building materials. There are also positive attributes in all building materials. With the proper methods for material evaluation and installation, almost all building materials can be good choices. Where no standards for evaluating such materials have been written, there is a need to prove your assumptions of material safety.

The way to gain acceptance for these types of materials is through structural engineering. The Code was structured, in part, based on these principles, so it is fitting that where no standards for testing or evaluating are established, basic engineering prevails.

The traditional structural analysis methods can be used to assert structural integrity. Most materials have certain structural properties that are known and can be used in formulae to create a rational analysis of a structural system. Others, such as straw bale construction, as illustrated in Figure 3-11, can be tested to the limits of destruction and the resisting forces established. The forces of gravity and wind are known, so it is a matter of applying those loads onto the framework to be considered and evaluating the result. There may be redundancy incorporated if an engineer performs the analysis and tends toward the conservative, designing it so that it is stronger than necessary. For example, added tension rods may be placed at strategic places into a framework of a natural wall framing system such as straw bale just to ensure that forces are resolved and that the allowable stresses of the material are not exceeded.

Full load testing is another analysis method open to the building official. This type of test is used less often because it is difficult to replicate all natural forces working in concert acting on a building. An example would be building out of recycled glass bottles, filled with

Photo Courtesy of David Eisenberg, Development Center for Appropriate Technology

Figure 3-11 ■ Straw bale construction, while not within the prescriptive provisions of the IRC, is still allowed based on performance provisions of the Code. The use of alternative materials has long been a means of evaluating the viability of a new or innovative building material.

sand, and stacked together, making a wall frame. The bottles are held in place by fusing the glass together. The method would be regarded as *green* because it uses a recycled, sustainable material. Keep in mind that an engineer would need to be able to analyze the potential stresses at the joint connection and analyze the interaction between sand with a fluctuating temperature different from glass if they have different rates of expansion. There would be many engineering questions to resolve, so it may be easier to build it and see if it works. In this case, a full load test might be in order. The building official may authorize a one-time-only license to build a prototype, apply a load, and see how it performs. If it works, it could find its way into the mainstream.

■ TALK TO YOUR INSPECTOR

You must learn how to communicate with your building inspector. Your goal is to have a successful project. Between you and the fulfillment of that goal are many obstacles, including financing, material selection, design, contract, and construction documents. Getting through the process of inspections, then to earn a certificate of occupancy is a big challenge. Although the inspector is responsible to verify safety considerations, your job is to build in that safety to meet the requirement. Communicating effectively with the inspector is essential to meeting that challenge.

Understanding the Code

The language of the Code may seem complex, but it is really simple. Anyone can propose a Code change for the International Codes according to a set procedure. The Code is written by material manufacturers, contractors, architects, engineers, and building officials. Many Code changes are a result of relaxing a formerly stringent requirement or adding clarifying language to improve ambiguous wording. On a few occasions, there is a need to tighten up a requirement as a result of reported injuries or illness caused by a weak provision. The Code, although a set of prescriptive requirements, serves as a guide for the end result of building safely.

Appeals

You may think that the building inspector's word is final. With due process considerations, this is certainly not the case. The inspector has a supervisor, the building official, who has the latitude of allowing for modifications to the Building Code and the use of alternative materials and methods of construction. If you disagree with the inspector's decision, you can ask the building official to issue a decision on the issue. If there is still resistance from this approach, there is yet another avenue of appeal. The building official has to decide appeals as well.

> *In order to hear and decide appeals of orders, decisions or determinations made by the building official relative to the application and interpretation of this code, there shall be and is hereby created a board of appeals. The building official shall be an ex officio member of said board but shall have no vote on any matter before the board. The board of appeals shall be appointed by the governing body and shall hold office at its pleasure. The board shall adopt rules of procedure for conducting its business, and shall render all decisions and findings in writing to the appellant with a duplicate copy to the building official.*[9]

Based on state law or local ordinance, any decision made by the building official when related to applying the Code may be appealed to this body. This is normally a volunteer group of builders, architects, engineers, and subcontractors who listen to both sides and render a decision for or against the building official's decision. The board of appeals' decision may be appealed further through the court system. In some states, the decision by the board of appeals is advisory. In others, it is a due process step toward legal appeal.

How to Communicate

Clear communication is the key to a successful collaboration when it comes to working with inspectors and building officials on your home building project. Following are a few points to remember for reaching a proper level of understanding:

- Learn about the Code that affects your construction. Know the proper questions to ask your inspector. Learn the language and demonstrate respect for the law that regulates construction.
- Keep your options open and know what to do if you and your inspector reach an impasse. The inspector has the responsibility to verify safety requirements for your construction project. An alternative approach that you propose may be acceptable or may have to be approved by the inspector's supervisor. Do not get mad at the inspector if he or she tells you that the plywood you installed is not marked with proper identification or is installed improperly. Search out the reasons and explore the possibilities for an alternate approach.

[9] *2006 International Residential Code, International Code Council, Inc., Section R112*

- View your inspector as a coach and a guide. Ask for advice and follow the inspector's suggestions. If something seems too complex, ask for an explanation. The inspector has seen this type of construction perhaps more times than you have and is loaded with all sorts of different approaches to construction and is a great source of advice.
- Use reason when arguing a point of contention about a Code issue. Do not rely on emotion to win an argument. Inspectors have no emotional tie to your project. They see your project as one of hundreds they are responsible for on a continuing basis.
- Be considerate of your inspector. Consider how your actions or manner affects him or her. Try to avoid actions that adversely affect your professional relationship. A cold soda on a hot day or a friendly handshake and a smile do a lot to improve your day-to-day interaction and develop a rapport.
- Respect the authority of your inspector. He or she can be your best ally if you get into trouble. The inspector is not likely to develop animosity unless you give him or her reason. Listen to the inspector's counsel and follow his or her advice.
- Be open to new ideas and concepts and new ways of doing things. Listen to the advice of your inspector, who is in a position to know the alternative ways of doing something that works.

Inspector

There may be times when you encounter an unreasonable inspector. Despite the education, experience, and professionalism characteristic of most construction professionals, there are some who have poor manners. It may be a personality conflict or differing backgrounds, but when conflict arises, follow these steps to extricate yourself from a sticky situation and take the high road:

1. Be calm and consider the consequences of your actions. Be polite. Conflicts originate in part when there are signs of intentional lack of respect and a visible display of emotional reaction.
2. Try to reach the inspector on an objective level. Use reasoning and facts. When people disagree, they connect emotionally to their position. Use logic and rational reason to demonstrate your position.
3. Ask specifically what things need to be corrected. When an inspector denies an approval, he or she should detail the corrections needed. You may get a laundry list, but at least you will have a thorough list that you can work on. This allows the two of you to focus on the issues and not on personalities.
4. Document your discussions and take notes for meetings with your recollection of agreements. You may need these when discussing the issues with the building official. This also ensures that subsequent arguments are based on initial points and not a new or never-ending list of corrections.
5. If all else fails, discuss your concerns with the inspector's supervisor, the building official. Try to use this option sparingly. Enlist the support of the inspector with a request: "I know that we agree that this installation looks okay, even though it is outside of the conventional way of building. Would you help me by telling your boss that the construction looks just fine?"
6. Barring resolution with the building official, the last informal step would be to take your case to the board of appeals.

Everyone agrees on the broad scope of safety and the mutual goal of building safety; we just disagree sometimes on the details. Reach out to understand. You are successful in your present profession because you do a lot of things right. Add the aspect of effectively communicating with your inspector and see how that improves your project.

■ GREEN BUILDING AND THE EVOLVING BUILDING CODE

Over the past century, model codes with prescriptive delineation of requirements marginalized the use of alternative materials commonly used in green building. In the last 15 years, building safety professionals have had a significant effect on the acceptance of green building materials and techniques. As the gatekeepers for acceptance of alternative materials and methods of construction, they have helped advance this emerging movement. Because of the significance of green building and sustainable design, the building safety community has embraced and continues to embrace the concept and become an advocate of this type of construction.

Impact of the Codes on Green Building

The ICC is taking an active role in responsible activities that are environmentally and economically sustainable as they relate to building safety. Green building and sustainable design proponents have already influenced the Codes. Energy efficiency has become an integral aspect of the building safety profession for almost 20 years. Though born out of the energy crises in the late 20th century, it found its way into the world of building safety as early as 1988 as the Model Energy Code. As a method of reaching out to embrace sustainable design, the Performance Code was developed in 2001. This Code's objectives include achieving intended outcomes for fire safety, structural integrity, energy-efficient materials, and plumbing systems. Moreover, efforts are underway to take the next steps forward in accepting principles of environmental design as mainstream instead of marginal. The ICC has entered into a partnership with several constituent groups, including the National Association of Home Builders (NAHB) and the United States Green Building Council, to further advance the cause of environmentally responsible building methods. This partnership will encourage the acceptance of green building practices. The intent is to remove stumbling blocks from the path toward sustainable design and green building as well as adding the tools that make it easier to meet these standards.

A guiding principle of the Codes has always been to encourage innovation among inventors while maintaining the minimum level of safety afforded by conventional construction materials and methods.

Greening of the Codes

Green home building has become more mainstream and standards are being developed. In response to the need for environmentally responsible values within the framework of the Building Code, the ICC is coordinating with the NAHB to create a standard for green homes.

An ICC/NAHB National Green Building Standard (NGBS) has been developed. The intent is to create an enforcement document for green home building criteria. This would focus on residential buildings. The standard has just been completed and has recently been accredited by the American National Standards Institute (ANSI) as an approved standard. This standard is the first green home building standard to attain this accreditation.

In addition, the International Code Council is developing a Green Construction Code that will detail substainable or green building criteria and provide a measurement for compliance. The Code should be available in 2010. For details, visit http://www.iccsafe.org.

BEFORE YOU DECIDE . . . REFLECTIONS AND CONSIDERATIONS

- ✔ Some manufacturers create building products that are substandard. Some builders make improper installations. Safety is the most important aspect when building a home and is regulated by many different laws.
  - The Building Code is just one of these laws.
  - Safety is the primary concern in building. The inspectors have an obligation to enforce the Codes that have been established to ensure this safety.
  - Invariably, conflicts will arise between what you want and what building safety regulations will allow.
- ✔ The Code includes two approaches to compliance alternatives: prescriptive provisions and performance provisions.
  - The prescriptive approach is like a cookbook approach with specific sizes, capacities, conditions, and so on. There is no allowance for deviation from this approach.
  - The performance approach has the end result in mind, but it also allows you to compare your approach with a baseline (like the prescriptive approach).
- ✔ Alternative building materials and the green building movement are relatively new.
  - New innovations are always improving a past process or material, allowing for greater efficiency and lower impact on the environment. As such, officials who apply these Codes struggle to keep up with advancing technology, always considering safety as paramount.
    - Some alternative materials are natural and renewable such as adobe or straw bale construction. Although not specified in the Code, these materials may still be allowed.
  - Modifications to the Code allow material usage even if not within the Code.
    - You must demonstrate that the new material or method meets the intent for safety.
- ✔ Communicate with your inspector in an honest and candid manner. An inspector should consider your request in a fair manner. There may be valid reasons for the inspector to question your plan.
  - An option is to appeal decisions to the building official or board of appeals.

✔ Green building is becoming more common in the realm of the building inspection.

- The Codes are becoming greener.
- There is more influence on alternative materials and methods of construction.

FOR MORE INFORMATION

International Accreditation Service (IAS)
http://www.iasonline.org/

International Evaluation Service (ICC-ES)
http://www.icc-es.org/

Chapter 4

RESOURCE EFFICIENT DESIGN AND MATERIAL USAGE

SMALL IS BETTER

A common thread in every culture is habitat. Constructing a home may represent the largest investment we have in our financial lives, but it also holds another designation. Building a home has a large and negative effect on the environment in several ways. The larger the house, the more materials are used and the more resources are expended. New house construction compounds into a detrimental effect on the environment.

Size Matters

Large houses consume more energy than small ones. They use more building materials, cover more natural grade (that would normally serve to replenish the water basin), and produce more construction debris. During the life of the home, more energy is consumed to heat and cool a larger home. In fact, the size of a home has a greater impact on the environment than any other single factor, including energy efficiency or human behavioral changes. According to the United States Census Bureau, the median size of an American home in 2006 was 2,248 square feet.[1]

The comparative price of housing increases as well. According to the National Association of Home Builders (NAHB), the average price per square foot in 2006 was around $92 up from $72 in the year 2000, a 28% increase in 6 years.[2] At that rate, the cost of the average new home is over $200,000. Because most home buyers finance their ownership of a home with a mortgage, they are limited in their purchase price based on their ability to repay the debt. Normally, buyers will purchase as much as they can afford, considering it to be a smart investment technique. Demand drives supply, and developers build to satisfy consumers. The economy has thus built a system that inherently brings harm to the environment: by building increasingly larger houses in a manner that does not consider the impact of construction. You could

[1] *U.S. Census Bureau, http://www.census.gov/*
[2] *National Association of Home Builders (NAHB), http://www.nahb.org/*

build a smaller green home and still include building technologies that improve thermal performance and energy efficiency. Because of this, your home will require a substantial investment. Rather than size, that investment will be based on the green characteristics of the home.

House Size and Your Needs

Finding just the right size house for your needs is not difficult. You need to have a functional home that allows growth as your life changes but within reason. Selecting the right size and design is much like controlling your appetite—exercise moderation and do not overindulge. Consider the theme of your home as being a green home. That implies *modest* rather than *opulent*. Ponder the following design questions regarding your lifestyle and family needs:

- How long do you expect to live in this house?
- Will your family entertain overnight or weekend guests?
- Do you desire a view between certain rooms?
- Do you like sun shining in certain rooms at certain times of the day?
- Do you want interior visibility of outdoor plants?
- What use is needed for each room?
    - Living room: Is there a need for a formal living room? Could a family room substitute?
    - Dining room: Is there a need for a formal dining room? Could a kitchen space substitute?
    - Bedroom: How many bedrooms are needed and how large do they need to be? How will they be used? Sleeping primarily? Office? Hobby?
    - Kitchen: How much cooking will you do? Will this be a gathering center A passageway? What appliances are expected?
    - Garage: What size garage is needed? Could it double as a workshop?
    - Bathrooms: How many bathrooms are needed?
    - Closets: How much closet space is necessary?

Only you can determine the absolute minimum size home you need for your lifestyle and that of those who will live with you. Figure 4-1 illustrates a comparison of house size to occupants. House size should reflect the amount of people living in it. There is little research available to help you because lifestyle choices vary. With the size of the average family steadily decreasing (it is now somewhere between three and four members), the average home size is increasing to well above 2,300 square feet. The Building Code could be used as a guide: Table 1004.1.2 establishes the anticipated occupants in a fully occupied building for the purpose of establishing required exits. Residential uses reflect 200 square feet per person as a proper ratio. Although this number seems low (an 800 square foot house may be a bit small for a family of four), it establishes a baseline. Somewhere between these high and low extremes is the size of a functional house that will fit your needs.

The NAHB has developed *Green Home Building Guidelines* that grant points for the size of the home based on the number of bedrooms. The point system increases as the size decreases for comparative bedroom units. It is a good place to start when considering the necessary size for your home.

The smaller the home compared to the number of bedrooms, the higher the point scale rises. By doing some math and considering your needs and your *shade of green*, you can work toward an appropriate size.

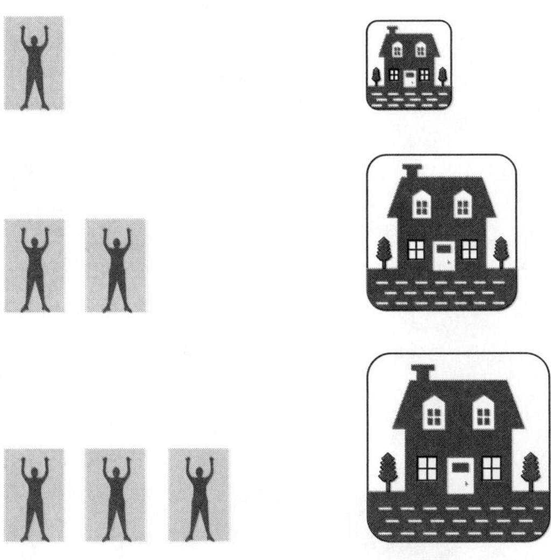

Courtesy of Lynn Underwood

Figure 4-1 ■ The size of the home should compare to the number of occupants. Work toward a more appropriate size based on the number of bedrooms and bathrooms needed. Consider if ancillary spaces such as a formal living room, formal dining room, parlor, or den are really needed.

CONSERVATION OF MATERIALS

The first step is to reduce the use, or at least the misuse, of all materials. Proper design and careful monitoring of the construction process will help achieve this goal. All construction involves some waste, most of which can be avoided with thought for conservation. Properly managing a project such as building a home and controlling job waste in this manner could mean always being on the site and watching what takes place. There are numerous other concepts that can be used to improve efficient use of new materials or the recycled use of existing materials.

Controlling Job Waste

Human activity includes production and consumption. Because we are imperfect, human activity also includes waste. The average per person disposal rate is 2.8 pounds of solid waste per day. The result of that waste includes a direct cost of the useful products being wasted and an indirect cost of storing that waste. The Environmental Protection Agency (EPA) estimates that 136 million tons of building-related construction and demolition debris was generated in the United States in 1996 alone.[3] Most of this waste comes from demolition and renovation. A portion comes from new construction. Approximately equal portions come from commercial and residential construction.

The most common materials to be discarded at a construction site include wood, drywall, plaster, bricks, roofing, and plastic. A typical waste disposal canister is illustrated in Figure 4-2. For building debris, the EPA estimates that the overall percentage of materials in construction and demolition debris falls within the ranges shown in Table 4-1.

Table 4-1 illustrates that concrete and wood represent the largest volume of construction materials disposed. Along with drywall, these three items represent around 90% of all waste. All of this waste must be disposed and stored in a landfill.

[3] *United States Environmental Protection Agency, www.epa.gov/*

Courtesy of iStock Photo

Figure 4-2 ■ Construction waste represents significant environmental damage as well as lost material resources. Control of that waste is essential for a green home.

■ Table 4-1

A variety of building materials are discarded during construction activities. The listed materials and their associated ratios illustrate volume of waste by comparison

Concrete and mixed rubble	40–50%
Wood	20–30%
Drywall	5–15%
Asphalt roofing	1–10%
Metals	1–5%
Bricks	1–5%
Plastics	1–5%

Source: Environmental Protection Agency[4]

Around 10 years ago, the NAHB estimated that a builder paid $511 per house just for the collection and disposal of construction waste.[5] Although the cost has undoubtedly increased, this construction waste amounts to 8,000 pounds or 50 cubic yards of waste for a 2,000 square foot house. This waste includes the various materials with respective amounts as listed in Table 4-1.

A properly maintained construction project includes methods and practices of controlling waste. Part of any contract with a builder should include your desires for waste management. One useful way to conserve building materials and avoid waste is to have a plan. Consider every way you can design and build with the least materials and produce the smallest amount of waste.

[4]*EPA, http://www.epa.gov/*

[5]*National Association of Home Builders (NAHB) (U.S.) Research Center (1996), Residential Construction Waste Management Demonstration and Evaluation, Task 2 Report*

Framing Cut List

A framing cut list plan provides step-by-step procedures for the cutting and layout of each piece of lumber. It is a catalog for material needed and a master plan for material usage. The cut list provides an accurate inventory needed for assembly as well as a take-off list for material purchase. The cut list can be derived from a detailed framing plan showing elements within a wall, floor, or roof layout. This type of framing detail plan depicts the location and size of various components of a wall frame such as headers, trimmer and king studs, cripples, essential blocking, and even top and bottom plates. It also shows floor and roof framing and the locations of all elements. For instance, it may denote that a roof truss will bear directly on top of a stud within a wall frame so that a second top plate can be omitted. A plan such as this will remove the use of unnecessary material routinely cut for such conditions as blocking, bridging, bracing, backing, or thermal barriers. Using a cut list and plan in this manner is a great way to reduce construction waste that would otherwise be realized without the plan. The cut list will illustrate the cuts you can make to framing lumber, avoiding duplication or similar underuse of a framing member. For example, if you need two pieces of 2 × 12 as header stock that is 33⅝ inches long and you had two boards to pick from, one 72 inches long and another 96 inches long, you would select the smallest that would yield the least waste. In this case, the 72-inch board would result in a small piece, 4¾ inches, as waste. Without a plan, you may cut the 96-inch board, resulting in a 29-inch board as waste.

Design with Standard Lengths

Framing lumber and sheet materials are manufactured or cut in 2-foot increments. Knowing this, design your building to match this industry standard. The result will be less waste and less work than making the cuts necessary to match your arbitrary dimensions. This option exists for floors, walls, roof, and ceiling.

Wall framing

During the design process, you can conserve waste in wall construction. Instead of a wall length of 31½ feet, use a 32-foot wall length and conserve lumber that would otherwise be wasted. You can use top and bottom plates in the 16-foot length and a minimum quantity of studs and exterior siding and interior finish wallboard. The same is true for the height of wall frames. If the intended ceiling will be 7 feet 6 inches, use a standard 8-foot height and conserve the lumber.

Floor framing

As with walls, floor framing can be adjusted to match 2-foot increments and save material. Because of the significantly larger material, the savings will exceed that of wall framing. The savings will extend to the surface of the framing as well. The width and length of sheet materials such as plywood decking is normally 4 feet by 8 feet. This panel size will align nicely with the dimensional shape if you maintain these dimensional limits.

Roof framing

Roof framing will follow the same principle of savings. Roof framing of solid sawn lumber, although smaller than floor framing, is certainly larger than the elements of wall framing. In addition, the savings on roof decking will be the same except that any pitch will change the overall latitudinal length. You can design this as well by factoring in the added length based on the desired pitch by using the Pythagorean Theorem. The square of the hypotenuse of a right

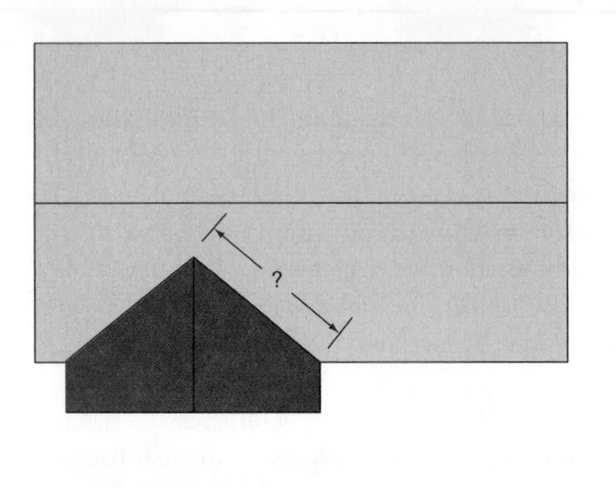

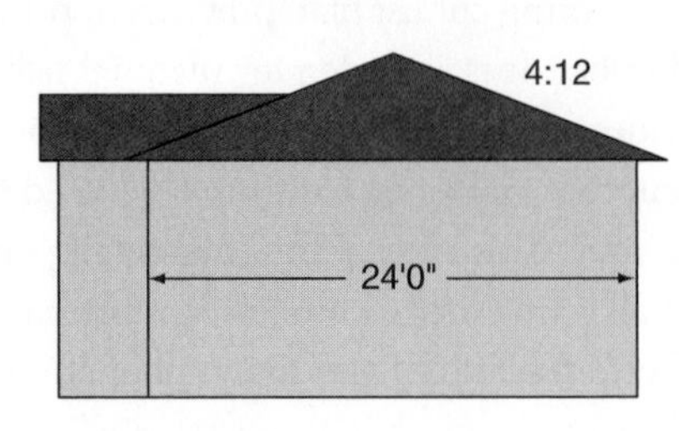

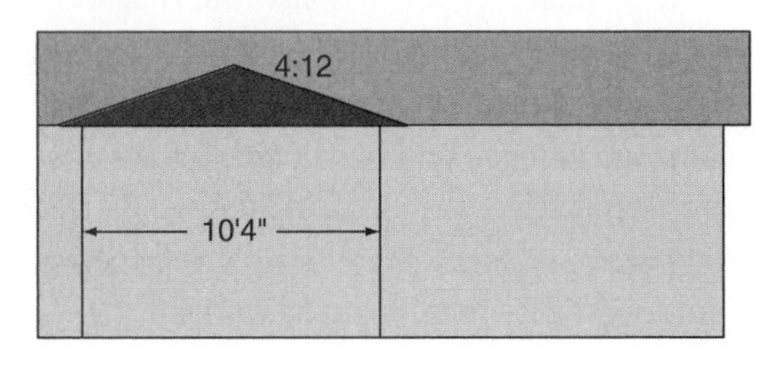

Courtesy of Lynn Underwood

Figure 4-3 ■ Math is essential to determine building size and to design to proper dimensions. Using trigonometry, you can determine the length of the hypotenuse of a right triangle if you know the other two dimensions (length and height). This will ensure a proper cut with minimal waste.

triangle is equal to the sum of the squares of the two sides. A 48-foot simple pitch roof with a 4:12 pitch is essentially two right triangles with the following dimensions:

$$48 \text{ feet}/2 = 24 \text{ feet (symmetrical rafter system)}$$

$$\text{Height of rafter} = (4/12) \times (24) = 8 \text{ feet high (in the center)}$$

$$\text{Hypotenuse calculation } \sqrt{(24^2) \times (8^2)} = \sqrt{640} = 25.3 \text{ feet (rafter size)}$$

Figure 4-3 illustrates a typical elevation view of a home with various pitches in the roof line. Remember, you still need to add length for overhang and determine the roof pitch needed to make the layout comport to the increments of panel size. Also, in the case where a second top plate is omitted, it is necessary to set the bearing portion of the roof framing directly over a bearing stud within a wall frame, so it will be essential to delineate this on a plan to ensure structural integrity.

Use of Alternative Framing Techniques

There are numerous techniques that you can use to realize savings. Some of them seem obvious. For instance, if you need to build a frame 6 feet long, cut 12-foot lumber in half instead of using 8-foot lengths and creating 2 feet of waste. Perhaps you could increase the center spacing for framing members to the Code allowance. Efficiency in framing will enable savings in other aspects of the construction process that follows. For example, if you use less framing lumber, you use less nails. Less concrete for a footing requires less digging. If you conserve at the front end of the project, the principle carries forward to the end. The following examples are intended to be a guide toward this objective.

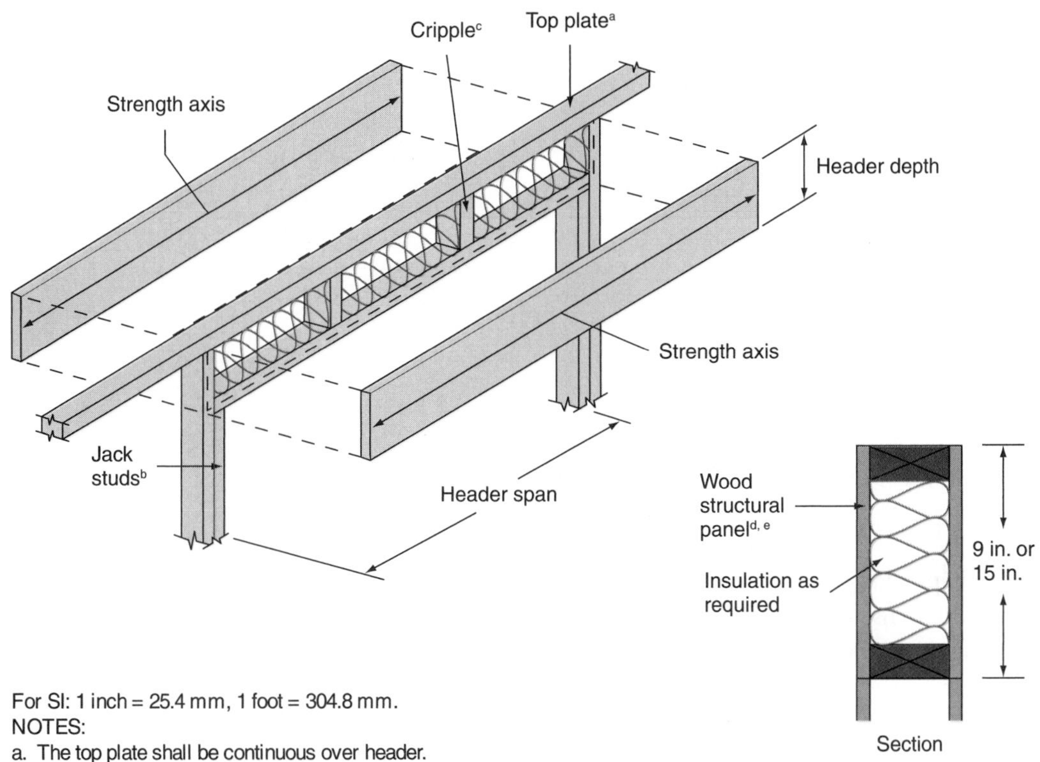

For SI: 1 inch = 25.4 mm, 1 foot = 304.8 mm.
NOTES:

a. The top plate shall be continuous over header.
b. Jack studs shall be used for spans over 4 feet.
c. Cripple spacing shall be the same as for studs.
d. Wood structural panel faces shall be single pieces of $^{15}/_{32}$ inch-thick Exposure 1 (exterior glue) or thicker, installed on the interior or exterior or both sides of the header.
e. Wood structural panel faces shall be nailed to framing and cripples with 8d common or galvanized box nails spaced 3 inches on center, staggering alternate nails ½ inch. Galvanized nails shall be hot-dipped or tumbled.

Reproduced with permission from the International Code Council, Inc.

Figure 4-4 ■ The use of the prescriptive design for a wood structural panel box header can reduce the material usage while still complying with requirements for a safe house. This detail makes use of a wood structural panel instead of solid sawn lumber.

Use of insulated box headers

A box header is an alternative that provides the same function as a solid sawn header but uses significantly less material. For example, a doubled 2-inch by 12-inch solid sawn header 8 feet long has about 16 board feet by itself. Considering the extra trimmer studs and cripples within the framework, there is even more. A box header, as depicted in Figure 4-4, uses a wood structural panel installed on the face of the wall frame, defined by a wood plate (2 inches by 4 inches), where the header would have been. This represents only 5 board feet of lumber—a 67% decrease in material.

As shown in Table 4-2, a wood structural panel box header within a wall frame can safely support loads based on floor or roof supporting conditions and house depth. Header spans up to 8 feet are permitted based on these loading conditions and panel depth.

Use of two-stud corner and wall intersection framing

The International Residential Code (IRC) requires only two studs in an outside corner with an appropriate wood backup cleat, metal drywall clip, or other approved method. These two studs would be one at the end of the intersecting wall. Blocking between studs in this position serves as the backup cleat to secure the intersecting wall rather than wasting a full height stud

■ Table 4-2

Maximum Spans for Wood Structural Panel Box Headers[a]

		House Depth (feet)				
Header Construction[b]	**Header Depth (inches)**	**24**	**26**	**28**	**30**	**32**
Wood structural panel—one side	9 15	4 5	4 5	3 4	3 3	— 3
Wood structural panel—both sides	9 15	7 8	5 8	5 7	4 7	3 6

For SI: 1 inch = 25.4 mm, 1 foot = 304.8 mm.

[a]Spans are based on single story with clear-span trussed roof or two-story with floor and roof supported by interior-bearing walls.

[b]See Figure 4-4 for construction details.

and still provide adequate support for the drywall. This way, you can make use of scrap lumber. In addition, use of an *approved* (listed and tested for that purpose) backup cleat performs the same function but with metal or plastic. These metal or plastic drywall clips, as depicted in Figure 4-6, are used in place of a third stud at a corner or an intersecting wall and a supporting shelf for the drywall's edge. Removing nonessential solid framing in floors, walls, or roof systems increases thermal efficiency by adding insulation. Two-stud corners with blocking or backup cleats are an option used to create an airtight connection to drywall.

Exposed Wood

Using interior exposed beams that require no additional covering such as drywall is a good way to conserve resources. Interior exposed wood can add beauty to your home and decreases the material used, but there are limitations. Section R315 of the IRC establishes a minimum flame spread and smoke developed threshold, which the material may not exceed. However, most species of larger wood beams or posts meet these criteria. In addition, wood beams in exterior locations may need to be protected from weather depending on the exposure. Check with the local building official on any particular questions.

Manufactured Products

Anytime a product is manufactured and not natural, the use is tightly regulated and limited according to the use intended by the manufacturer. Use these manufactured products of systems as they are intended. There are many varieties of innovative products that are available in the construction marketplace. Chapter Five more thoroughly develops the use of these innovations in building materials. Panelized wall, floor and roofing assemblies, engineered lumber, trusses, and similar modern developments all bring efficiency to a profession accustomed to field assembly of all parts. The following elements of construction all represent manufactured innovations that would help achieve your objective of a green home:

- Structural insulated panels (SIPs) are available for walls, floors, or roofs.
- Engineered lumber, such as glulams, structural composite lumber, or finger joint studs are available materials.
- Factory-framed floors, walls, and roof systems bring manufacturing precision to your job site and all but eliminate waste.

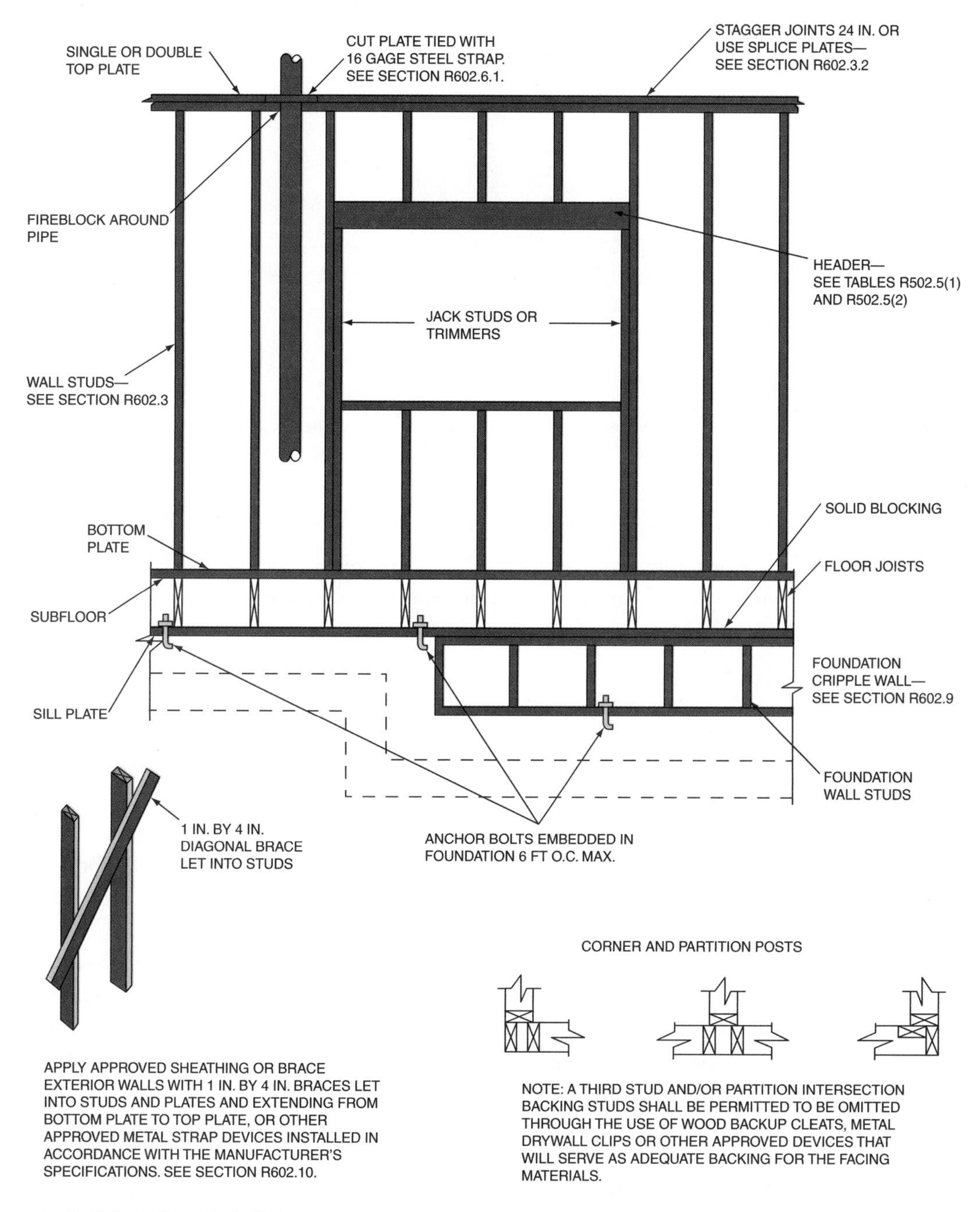

Source: Reproduced with permission from the International Residential Code 2006, Country Club Hills, IL: International Code Council, 2006, Figure R602.3(2)

Figure 4-5 ■ An intersecting wall frame at a corner or partition post normally requires three studs to make a secure connection. This detail illustrates a means of reducing the number of studs to two under certain conditions. Note that Figure 4-5, extracted from the IRC, illustrates that this is an approved method for constructing a wall frame.

- Engineered trusses are a designed mechanism that relies on comparatively slender members connected at joints to transfer loads to a point of support, normally each end.
- Plywood is manufactured using wood veneer that is cross-laminated and sealed with heat, pressure, and an adhesive appropriate to the application.
- Oriented strand board (commonly abbreviated OSB) is manufactured from wood fibers with heat-cured adhesives.

Courtesy of Building Energy Resource Center, http://resourcecenter.pal.gov/[6]

Figure 4-6 ■ These drywall clips are used in place of a third stud at a corner or an intersecting wall.

- Precast aerated autoclaved masonry is used in making wall units.
- Engineered exterior siding and trim are available in numerous options, including composite material, vinyl, plastic, hardboard, Masonite, exterior insulated finish systems (EIFS), stucco products, and fiber cement composites.

Minimum Standards

The use of *minimum* standards, where appropriate, offers conservation of materials while still providing the necessary safety. Although the Code is a minimum document, setting out the baseline for the elements essential to a building, it is based in science and the weather-related geographic conditions that are expected during the lifetime of the structure. Natural forces, such as wind loads, seismic intensity and frequency, rainfall amounts, snow load, heat, and cold all form the basis for this minimum standard. For example, whereas you could reasonably expect to build a mostly flat roof in Arizona because the prevailing rainfall is small, it would not be the right thing to do in Oregon. A building on the outer banks of North Carolina would have more shear wall bracing to resist hurricanes than one in New Mexico that has a much lower prevailing wind speed. By adding these elements in an area where they are not required to resist natural forces, the building uses additional material resources unnecessarily. Minimum Code standards are adequate. The following are just a few of these minimum standards that appear in the Code. Their use in construction represents a low impact on the environment and conservation of natural resources.

Foundation material conservation

Use minimum size for concrete footings and foundations. The IRC establishes the minimum footing size for various loading conditions identified by floors and the weight of the bearing walls. The size is based on the allowable soil bearing value and the comparative weight of the bearing walls (conventional light frame, hollow or solid masonry).

Wall material conservation

Use minimum size beams, headers, and studs and no load-bearing headers where not required. There are tables within the IRC that set out the minimum requirements for headers and girders on both exterior and interior bearing walls. Use these minimum sizes instead of using doubled 2-inch by 10-inch or 2-inch by 12-inch built-up headers, which has become common practice. Additionally, do not use headers in interior wall frames that are non-load bearing. According to the IRC:

Nonbearing walls. Load-bearing headers are not required in interior or exterior nonbearing walls. A single flat 2-inch-by-4-inch (51 mm by 102 mm) member may be used as a header in interior or exterior nonbearing walls for openings up to 8 feet (2438 mm) in width if the vertical distance to the parallel nailing surface above is not more than 24 inches (610 mm). For such nonbearing headers, no cripples or blocking are required above the header.[6]

Top plate

The IRC permits the use of a single top plate (as opposed to two top plates).

Top plate. Wood stud walls shall be capped with a double top plate installed to provide overlapping at corners and intersections with bearing partitions. End joints in top plates shall be offset at least 24 inches (610 mm). Joints in plates need not occur over studs. Plates shall be not less than 2-inches (51 mm) nominal thickness and have a width at least equal to the width of the studs.

Exception: A single top plate may be installed in stud walls, provided the plate is adequately tied at joints, corners and intersecting walls by a minimum 3-inch-by-6-inch by a 0.036-inch-thick (76 mm by 152 mm by 0.914 mm) galvanized steel plate that is nailed to each wall or segment of wall by six 8d nails on each side, provided the rafters or joists are centered over the studs with a tolerance of no more than 1 inch (25 mm). The top plate may be omitted over lintels that are adequately tied to adjacent wall sections with steel plates or equivalent as previously described.[7]

In order to obtain an 8-foot interior finish ceiling height, most builders use double top plates with 92 5/8-inch precut studs. If you switch to a single top plate, you will need to buy 96-inch studs and cut them to 94 1/8-inch lengths.

Use maximum center spacing for framing elements

The IRC sets out the maximum center spacing for wood studs within a wall frame. Take advantage of this maximum and save as much as 25% on studs framing a wall. Center spacing of 19.2-inch or 24-inch center spacing could be used in many cases instead of 16-inch on center (O.C.) for floor systems, bearing wood stud walls, roof systems, or interior partitions. In order to increase the center spacing, floor sheathing must be rated for the increased span. Floor (and roof) sheathing span rating is normally stamped on one side.

Use 2-inch by 3-inch studs for interior framing of non-load-bearing walls

The Code permits the use of smaller framing for wall frames that hold only drywall loads.

Interior nonbearing walls. Interior nonbearing walls shall be permitted to be constructed with 2-inch-by-3-inch (51 mm by 76 mm) studs spaced 24 inches (610 mm) on center or, when not part of a braced wall line, 2-inch-by-4-inch (51 mm by 102 mm) flat studs spaced at 16 inches (406 mm) on center. Interior nonbearing walls shall be capped with at least a single top plate. Interior nonbearing walls shall be fire blocked in accordance with Section R602.8.[8]

[6] *2006 International Residential Code, International Code Council, Inc. (2006), Section R602.7.2*
[7] *2006 International Residential Code, International Code Council, Inc. (2006), Section R602.3.2*
[8] *2006 International Residential Code, International Code Council, Inc. Section R602.5*

NOTE

Most manufactured window jambs are based on a 3½-inch stud wall dimension. Use of different framing materials or sizes will need to accommodate any windows and doors.

BEFORE YOU DECIDE . . . REFLECTIONS AND CONSIDERATIONS

✔ The size of your new green home should reflect your minimal needs.
- The more you build, the more impact is placed on the environment.
- The more debris and waste created, the greater the harm is to the environment through the construction process.
- Larger houses create greater footprints on the soil.
- The larger the house, the more utilities are needed for heating, cooling, and lighting.
- The larger the house, the more repairs, maintenance, and attention it will demand in its lifetime.

✔ Another significant aspect of a green home building is the avoidance of waste.
- Debris and trash from construction represents a large portion of landfill waste.
- Avoiding waste saves in two ways: less to purchase and less to impact the Earth.

✔ There are several ways of minimizing wasteful practices that are used in green building projects.
- The amount and type of material used in a home, more than any other single contribution, will identify it as a green home.
- Material can affect the energy efficiency of the heating and cooling necessary, thus reducing the impact.

✔ Using the least amount of materials to produce a safe house is another hallmark of a green home.
- The Code allows for minimum material. Think smarter and use common sense when building.
- Use alternative construction techniques that are approved.
- Have a plan for material usage.

FOR MORE INFORMATION

Building Energy Resource Center
http://resourcecenter.pnl.gov/

Environmental Protection Agency (EPA)
http://www.epa.gov/

National Association of Home Builders (NAHB)
http://www.nahb.org/

U. S. Census Bureau
http://www.nahb.org/

Chapter 5

BUILDING MATERIAL CHOICES

CONVENTIONAL BUILDING MATERIALS

Even home builders can fall into a pattern of conditioned thinking. This pattern forms the basis for using conventional building materials or methods. It is rationalized that conventional materials are best for new builds because we know and trust them. We have the tools for them, whereas a new material would require an entirely new set of equipment. Conventional building materials work and, more importantly, the mortgage company trusts them enough to lend money to pay for them.

However, meaningful progress in life is the result of our capability to adapt. We need to adapt to change by meeting the new challenges that life presents and by adjusting to new technology sufficiently to survive and flourish. All of the positive changes in life are the result of considering and trying an alternative method. The use of conventional building materials does not negate your objective of building a green home. A green home can still be built by using conventional resources. There are pros and cons for using conventional building materials to build your new green home.

Conventional Wood Construction

Wood is currently a common building material in the developed world. Lumber is normally derived from trees and used for construction purposes when cut into shapes that are usable in the construction process. These cut pieces are available as boards, posts, studs, plates, joists, rafters, and beams. As a material, wood can be used to build just about any type of structure in moderate climates or even in areas with brutal natural forces. Wood has flexural and axial strength when supporting loads. It has strength in different loading conditions, such as a horizontal beam or joist, or in a vertical position, such as a post or stud. Wood products have different strengths depending on the species, varieties, or the location of the tree, as it is milled and the conditions during its growth. Before modern saws, a tree was cut to the needed length to serve its purpose, perhaps for use as a roof frame or in walls such as a log cabin. With modern equipment such as sawmills, trees are cut to serve specific uses that have become conventional such as wood beams, posts, and studs.

Wood Framing

Wood frame structures are built with a wood joist floor frame, wood stud wall frame, and a system of wood rafter roof frame. Typical wood wall frames such as those being erected in Figure 5-1 are

Photos by Lynn Underwood

Figure 5-1 ■ Erection of a wall frame of any height is a difficult task. When the wall frame is very tall, assistance is essential. Temporarily securing the frame is the first step toward final installation.

composed of nominal size wood pieces within the frames, such as 2 × 4 inch or 2 × 6 inch at the appropriate length. *Nominal dimension* is a term used for finished lumber sizes that are planed or milled to be a uniform size. The actual size is less than the stated nominal dimension. Table 5-1 illustrates the difference between nominal size and the actual lumber size for various shapes and sizes.

The wood frames (floor, wall, and roof frame) typically are covered on both sides with siding or sheathing as depicted in Figures 5-2 and 5-3. The outside is made to be weather resistant with siding or decking. The interior wall frame of the building has drywall or wallboard as sheathing. These sheathing materials develop stiffness in the frame, allowing it to resist more loads than it could not otherwise withstand. Specific rules for the assembly of these types of frames are found in Chapters Five, Six, and Eight of the International Residential Code (IRC). The advantages of wood framing include ease of installation and structural integrity when built properly. The

■ Table 5-1

Softwood Dimensional Lumber Sizes

Nominal	Actual	Nominal	Actual
1 × 2	¾ × 1½ in.	2 × 2	1½ × 1½ in.
1 × 3	¾ × 2½ in.	2 × 3	1½ × 2½ in.
1 × 4	¾ × 3½ in.	2 × 4	1½ × 3½ in.
1 × 6	¾ × 5½ in.	2 × 6	1½ × 5½ in.
1 × 8	¾ × 7¼ in.	2 × 8	1½ × 7¼ in.
1 × 10	¾ × 9¼ in.	2 × 10	1½ × 9¼ in.
1 × 12	¾ × 11¼ in.	2 × 12	1½ × 11¼ in.
3 × 4	2½ × 3½ in.	2 × 14	1½ × 13¼ in.
4 × 4	3½ × 3½ in.	6 × 6	5½ × 5½ in.
4 × 6	3½ × 5½ in.	8 × 8	7¼ × 7¼ in.

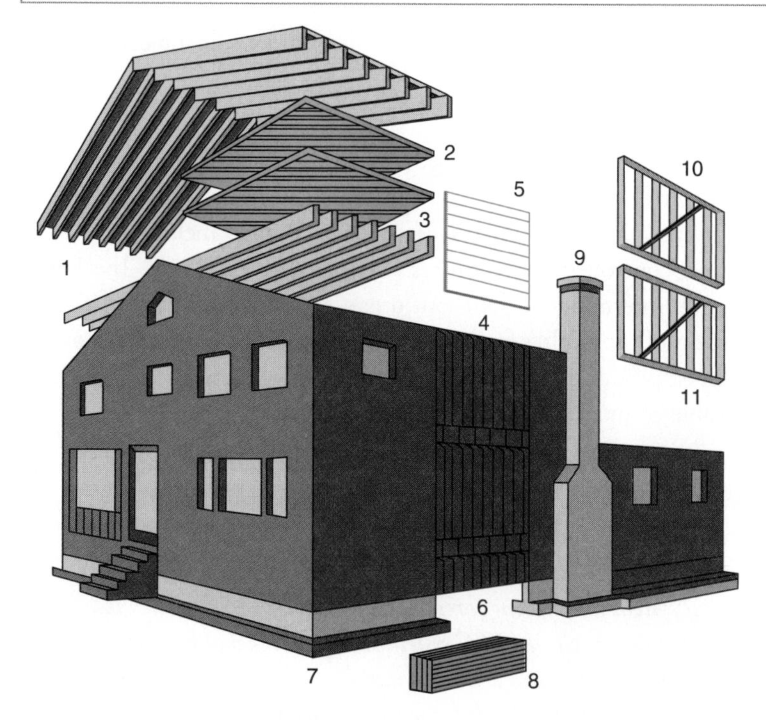

Lynn Underwood

Figure 5-2 ■ Various elements of a wood frame home include floor joists, stud wall frame, and rafter frame, along with brace wall panels, siding and floor, and roof decking. Generally, wood framing is supported on a foundation of concrete or masonry.

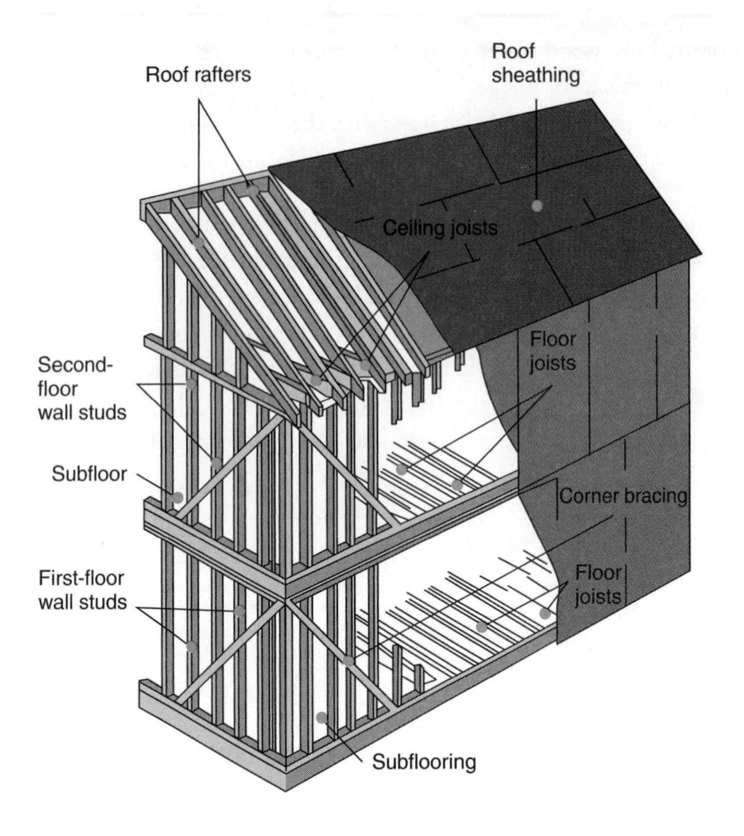

Lynn Underwood

Figure 5-3 ■ A wood frame building with identified elements such as the subfloor, studs, joists, rafters, ceiling joists, let-in bracing, shear wall panels, and sheathing.

disadvantages include long-term maintenance, if subject to insect infestation (termites); susceptibility to water or moisture damage; and unpredictability in strength and material quality across tree species. Wood framing, if not protected from weather or soil conditions, is subject to decay.

Use wood from sources that protect natural forests

Sometimes trees are harvested with little or no consideration for the environment. Clear cutting trees in a forest is a practice of harvesting trees in which virtually no trees remain in a sector. This type of harvesting adversely affects natural conditions, including erosion. Using lumber from trees within controlled harvesting is a responsible approach. The Forest Stewardship Council (FSC) is an international nonprofit organization established in 1993 to promote responsible management of the world's forests. This group sets standards and provides certification and labeling of forest products. These labels allow individuals to choose products from environmentally responsible companies that harvest natural wood.

> *The Forest Stewardship Council was created to change the dialogue about and the practice of sustainable forestry worldwide. This impressive goal has in many ways been achieved, yet there is more work to be done. FSC sets forth principles, criteria, and standards that span economic, social, and environmental concerns. The FSC standards represent the world's strongest system for guiding forest management toward sustainable outcomes. Like the forestry profession itself, the FSC system includes stakeholders with a diverse array of perspectives on what represents a well-managed and sustainable forest. While the discussion continues, the FSC standards for forest management have now been applied in over 81 countries around the world.*[1]

[1]*Forest Stewardship Council, http://www.fscus.org/*

Currently FSC-certified materials are being specified by registered design professionals as a commitment toward green building. The Leadership in Energy and Environmental Design (LEED) offers points or credits for using wood certified in accordance with the FSC principles and criteria for wood-building components.

Timber Frame

Timber frame is the process of constructing a building by using large sizes of wood hewn from trees. Normally these are in the form of beams and posts that are connected together through joinery at the ends of members. These techniques date back thousands of years. Modern use of joinery techniques has made simple connections associated with this type of construction, as shown in Figure 5-4. Notice the metal connectors that connect various elements. The benefits to this type of construction are structural integrity, higher fire resistance, and appearance.

Masonry

Masonry is one of the oldest building materials, having its roots in unreinforced earthen block. Masonry involves the laying of block units in a wall frame that are held together with mortar and sometimes filled with grout (if hollow). Masonry is used to build fences, monuments,

Courtesy of iStock Photo

Figure 5-4 ■ An example of a timber-framed homebuilding. The structural elements are larger posts, beams, and girders. Connections are typically bolted-type joints.

signs, and walls. It is versatile in that the blocks can be installed, cut, and shaped to create a pattern or defined shape. Its advantages include fire resistance, structural integrity, low maintenance, and relative ease of assembly (with experience). A disadvantage is that unreinforced masonry is susceptible to failure in a seismic event because of the weight.

Concrete

Concrete construction has numerous advantages, including structural integrity, low maintenance, high quality, and durability. Considering the durability and longevity of this material, concrete has some green aspects. Its ability to resist natural forces allows a structure built of concrete to be very durable and avoid the need to replace or rebuild. In addition, some innovations such as *insulated concrete forms,* or ICFs, make concrete even more green than other materials.

However, from the perspective of a green home, the use of concrete in any portion presents some harmful characteristics, including the following:

- The energy and heat required to synthesize Portland cement is enormous and normally requires burning fossil fuel with by-products.
- By-products of cement manufacturing include harmful volatile organic compounds (VOCs) such as sulfur oxides (including sulfur dioxide, carbon dioxide, oxides of nitrogen, and a host of similar resultant materials). In addition, solid waste such as clinker dust is created, although the industry normally recycles most of the material.
- During the mining process, rock and stone are taken from native conditions in watersheds, causing environmental damage.
- Transportation of cement to suppliers and the delivery of mixed concrete to the job site are necessary.
- Water consumption is required for mixing concrete.

Of particular concern where the use of concrete is considered is using Portland cement because it makes the most demands on the environment in energy drain and environmental damage. An alternative admixture is coal fly ash. Adding this fly ash to replace even a portion of Portland cement in some quantities is regarded as a green choice. According to Bruce King, PE, fly ash is

> *. . . an abundant industrial waste product that happens to be high in reactive silica, and thus an excellent pozzolan. For this simple reason it is rapidly becoming a common ingredient in concrete all over the world; it is already present to some degree in half the concrete poured in the US. Of particular interest to the industry is the idea of not just adding fly ash to known concrete mixes, but using large quantities to replace 30%, 50%, or more of the Portland Cement—the glue—in a concrete mix. Most of the reasons for using fly ash in any proportion are practical, such as increasing strength and durability, decreasing heat of hydration, and decreasing permeability. Those reasons alone make the idea of high fly ash concrete (HFAC) worth considering, but there are many global economic, health, and environmental concerns that make HFAC even more attractive and compelling.*[2]

Although concrete does have some negative attributes, its long-term durability in some ways compensates for its negative effects. Even with its environmental burdens, concrete has the benefit of an extremely long service life and resistance to natural decay compared to other building materials.

[2]*Bruce King, PE. (2007).* Making Better Concrete: *Guidelines to Using Fly Ash for Higher Quality, Eco-Friendly Structures. Green Building Press.*

USED, RECYCLED, OR RECLAIMED BUILDING MATERIALS

Used, recycled, or reclaimed materials are now regarded as completely acceptable for construction. In fact, using used or recycled materials is an affordable method of construction that is very environmentally friendly. The Building Code allows the use of used materials.

Approved materials and equipment. Materials, equipment and devices approved by the building official shall be constructed and installed in accordance with such approval. R104.9.1 Used materials and equipment. Used materials, equipment and devices shall not be reused unless approved by the building official.[3]

Although it is written in the negative, the approval by the building official as a condition of use is the key. The building official has the authority to allow reuse of material that has acceptable strength and durability. The test may simply be a matter of a building inspector's appraisal of the materials' condition, knowing the intended and immediate use. On the other hand, a tough analysis may be a professional's opinion on the condition of each component being considered. You may need to ask a professional engineer or general contractor to evaluate the used material and write an opinion for the building official. In any case, the approval will be predicated on the material's effectiveness for the intended use.

Reusing building materials is the finest form of recycling. It creates jobs as well. Numerous sources confirm that recycling creates many times the jobs for every one that would be created by land filling the same amount of waste. Those who proudly extol their recycling efforts often do so using the Mobius Loop illustrated in Figure 5-5.

Recycled Concrete as Aggregate

Concrete is widely available, is low in cost, and is relatively easy to work with. Concrete recycling is a significant contribution to a green home. According to the Portland Cement Association, 38 states currently use crushed concrete as a base for pavements such as streets and parking lots. Recycled concrete aggregate has passed the same tests as conventional aggregate. Concrete used in this manner is simply crushed into aggregate of a specific size and reused. There are other uses for this material as well. It can be used as a noise barrier or embankment, as underlayment under paving (even permeable paving), or for conditions where land needs fill material (although if supporting a structure, this may need a geotechnical design). In fact, the American Concrete Institute has written a standard for removal and reuse of hardened concrete.

There are other uses for recycled concrete that do not include the energy-intensive crushing and sorting. For instance, broken portions of slabs may be used as irregular masonry units for retaining walls or landscaping with the potential for groundwater recharge between the sections. If solid and intact, some sections of floor or wall slabs could be reused as wall sections of a new structure.

Reclaimed Flooring

Reclaimed flooring is a valuable source of high-quality wood or tile products. Much of the reclaimed wood comes from old growth timber and is a better wood than that from newer

[3] *2006 International Residential Code, International Code Council, Inc., Section R104.9*

Courtesy of photos.com

Figure 5-5 ■ The Mobius Loop has become a universally recognized symbol of the process of transforming waste materials into useful resources.

timber, which is manufactured from predominately younger trees. Having been installed as flooring, this wood has been pampered, cared for, and is in perfect condition. This wood can be reborn into newly installed flooring or in other trim uses such as baseboard, wainscoting, window frame, or even ceiling trim. With cleaning and new mortar, masonry tile can be used as interior flooring again.

Reusing Wood

Wood in any installation can find new life as long as its strength and size are adequate. Wood used in a structural system such as floor, wall, or roof frames are perfect for reuse. These frames have normally been covered during the life of the structure with sheathing or wallboard. Many times, the age is hardly noticeable and the quality of wood in older buildings often seems as good as new and sometimes even better. In addition, the cost is often significantly less than new material, especially if you manage to help perform the de-construction work in return for the wood.

Recycled (or Repurposed) Carpet

Floor covering, such as carpet and carpet pad, is recyclable and can be transformed into other products and appear new. Recycled carpet is normally identified as such by the carpet reseller. At the very least, select carpet that can be recycled. Note that recycled carpet disassembles the components of carpeting for remanufacturing. Because the materials are reconstructed, there is little cause for alarm regarding dust mites or other allergens. Recycled carpet is different from repurposing carpet in that in repurposing carpet, the carpet is reused in another building. Carpet that is relatively new or substantially clean could easily be repurposed and reused in another home. However, there are several drawbacks to repurposing, one of which is the health hazards presented by the possibility of relocating pests or other vermin. The green attributes include using a product that would otherwise be taken to a landfill and the corresponding costs of a new product, which include the embodied energy costs as well as the off-gassing of VOCs of new carpet.

Used and Recycled Roofing Materials

Roofing materials can also be readily recycled into a new product. Selecting this type of product supports the concept of recycling and encourages this initiative. In addition, most of this type of roofing is recyclable. Also, these recycled products are manufactured to the same specifications and quality as a similar new product that carries a similar warranty.

Examples of roofing materials that are appropriate for reuse include concrete tile, slate, and some types of metal. Many recycled roofing materials are available, including reinforced vinyl, rubber, and cellulose fiber.

Recycled Insulation

Insulation is a very common product that is available recycled material. Insulation is a product that retards the flow of heat through convective transfer. When a material is able to slow the movement of air to a standstill, heat flow in that manner stops as well. For instance, a wall cavity filled with insulation and sealed on both sides (with sheathing and wallboard) has very little convective heat transfer. Although the most common insulation materials are fiberglass and cellulose, many other products have a good track record as well. Cotton fiber in the form of clothing is recyclable and is an insulation material. Paper and cellulose are other materials that are commonly recycled into insulation material. Selections of any of these products that are reused represent a green project.

Recycled Porch and Deck Materials

Outdoor decking and porch materials are also items that take advantage of recycling or reuse. Many composite or plastic materials used in outdoor decking or porch construction are derived from recycled materials. Considering the alternative of filling a waste site such as a landfill, support of recycled products like this is certainly green. These products are engineered and manufactured to the standards for a similar newly constructed product, but waste is avoided. Most of these products have the added benefit of being recyclable when their useful life in your home ends.

Recycled Plumbing

Plumbing materials can be a source of reuse if caution is exercised. Salvaging a sink or bathtub might be valuable, but reusing a high water demand toilet (3-gallon flush) would be counterproductive. Water heaters should not be reused, particularly if the container is compromised through rust or decay. Pipes should be avoided unless they served as venting and were not subject to damage. Valves and controls (except for faucets in good condition) are not good candidates for reuse.

NATURAL OR RENEWABLE MATERIALS

Selection of natural content or renewable materials allows the use of nontoxic construction resources. But there is always a concern that natural materials decay faster than the manufactured type. This is where the life-cycle analysis (LCA) of a particular material is valuable.

> *Environmental Life-Cycle Assessment (LCA) provides a framework, approach, and methods for identifying and evaluating environmental burdens associated with the life cycles of materials and services, from cradle-to-grave.*[4]

The *Environmental Building News* further expands on what the LCA entails.

> *If you wanted to know about all the environmental impacts of a product, common sense suggests that you would have to trace that product from the origins of its raw materials, through its manufacture and use, and finally to its fate at the end of its useful life. That's the premise behind environmental life-cycle assessment (LCA)—a science that aims to quantify all the impacts of a product or service. LCA is often used by manufacturers to compare alternative ingredients and processes for making a product, and by policymakers for establishing preferences for one product over another.*[5]

LCA examines the environmental impact of a building material through its life before, during, and after construction. The environmental effect on the material is considered during manufacturing, packaging, and distribution, as well as what happens after its useful life. With this analysis, a true picture of the environmental friendliness of the material can be rated and compared to others. Because an LCA of a manufactured product will inherently have a more harmful effect on the environment, a natural content material will have significantly less harmful effect because the energy demand for manufacturing, processing, packaging, shipping, selling, and delivery would all be avoided. This is especially the case for one with an on-site manufacture like adobe, rammed earth, or straw bale. Another reason for using natural content material is that it is almost always a renewable resource.

Bamboo

Bamboo, like that illustrated in Figure 5-6, is a perennial evergreen plant that produces a wood-like material and is harvested as commercial timber. It is one of the fastest growing plants on the planet. It also has several other uses, including scaffold assembly, poles, beams, joists, rafters, and fencing. In fact, an entire house can be built of bamboo like the one illustrated in Figure 5-7. The tensile strength (ability to resist tearing apart) is near that of steel. It is a resilient and very flexible plant that rarely snaps in the wind. However, like wood, it is subject to deterioration and if left to natural forces, its strength can be eroded over years of exposure, leading to increased demands for repair and maintenance. Chemical treatment may be applied to prevent insect scavenging or decay. This is similar to the pressure treating of wood to resist termites.

Straw Bale

Straw bale construction has been around almost as long as straw bales have been stacked in barns. For well over 100 years in the Plains states, straw bale has been used to build barns, outbuildings, and even farm houses. Stacking the bales to form walls is the elemental method of a truly

[4] *Environmental Protection Agency (EPA), http://www.epa.gov/*

[5] *Environmental Building News, June 2008 issue No. 17:6, BackPage Primer—Life-Cycle Assessment: Tracing a Product's Impacts*

Courtesy of iStock Photo

Figure 5-6 ■ Bamboo harvested for industrial use in production.

Courtesy of iStock Photo

Figure 5-7 ■ As a sustainable building material, bamboo can be used to construct entire homes.

sustainable design and green building. Straw bale is a product that is organically inert and thus not attractive to insects. Its original manufacture as baled hay was intended to consolidate an organic feed material for livestock, holding the chaff together for storage until feeding. Stacking the hay provided protection from the elements and this led to the idea to use it to build a structure. The result was a habitat that resisted the cold winters and warm summers with thick insulation produced by a 24-inch-wide wall that was packed tightly with natural cellulose. The wall frame stood with little or no support except the friction between bales. The exterior was parged with a clay mixture to resemble stucco. This prevented weather infiltration along the sealed joints.

There are essentially two major types of designs for straw bale wall construction: bearing and nonbearing. The bearing wall type fully supports the roof load and the nonbearing type allows for straw bales to be placed within a load-bearing wall system, normally a post and beam wood structure (Figure 5-8).

Cork

Cork is a natural product harvested from the outer bark of a cork oak tree, as seen in Figure 5-9. Cork has a low specific weight, is partly elastic, and is flexible and durable. It is impermeable

Courtesy of iStock Photo

Figure 5-8 ■ Straw bale construction generally includes straw bale assembly and conventional wood frame roof assembly.

Courtesy of iStock Photo

Figure 5-9 ■ Cork is a sustainable building material because it is a renewable resource. Cork is used for a variety of building materials, including flooring.

to liquids (it floats and is widely used as a wine stopper), retards the flow of air (serves as insulation), and has a natural resistance to wear and tear. It has a resistance to fire and is stable dimensionally. In addition, it has a natural resistance to decay. It is grown mainly in Portugal and Spain and is harvested after repeated 9-year cycles of growth. The tree has a life span of around 200 years and generates several cork harvests in its lifetime. The bark is stripped, allowing new growth to take its place without harming the tree.

Cork has numerous uses in construction because of its mechanical properties. It is widely used in flooring, wall covering, and insulation. It is used in numerous other construction-related applications such as heat insulation, thermal isolation, flooring insulation, as well as a sound barrier and joint filler. It is totally renewable and its harvest does not harm the tree.

Earthen Materials: Adobe and Rammed Earth

Adobe refers to earthen construction; that is, using elements of the soil to form the elements of a structure. An adobe house is one where the walls are built with unburned sun-dried blocks of clay, straw, and other earthen materials. Sometimes the block is turned into a low-grade brick by exposing it to fire. These are called burned adobe bricks. These blocks or bricks are laid together much like blocks with a bonding pattern. The blocks are normally bonded together with a mortar of similar material. Adobe is regarded as unreinforced masonry (unless special reinforcement is added). When the wall reaches the final height, a concrete bond beam is added at the top. This bond beam is continuous around the walls, tying the system together.

Adobe, although much like masonry, has thermal resistant properties that retard heat flow. Because of the traditional wall thicknesses, heat is maintained throughout most of the daily heating cycle. Adobe is regarded as a renewable resource for building materials because when demolished the material reverts to the original constituent parts of soil. Adobe is regarded as one of the first building materials. It has been used since well before recorded history. Adobe construction is durable to most natural forces except earthquakes and flooding. Variations of the adobe wall include *rammed earth construction*. This method of consolidating and compacting adobe material between wall forms creates a wall that is very dense and structurally like concrete. In some cases, additives to the soil increase the resistance to water permeability.

Earthen materials may be used for floor surfaces as well to avoid using concrete, which is common. Earthen flooring is natural earth that is compacted with straw or other fibers and stabilized with various admixtures to form high-quality flooring. This type of floor can be attractive and comfortable. It can be made from natural renewable materials at a very low cost. The earthen floor is very durable with proper care. It requires little maintenance by the homeowner. Because of the low impact on the environment, renewable natural earthen materials such as adobe and rammed earth are regarded as truly green materials.

MODERN INNOVATIONS

Progress is the hallmark of any society. We grow in knowledge that we gain through experience. We use that knowledge to advance in technology, creating a better way of life. Such is the pattern of our progress. Construction follows the trend toward innovation. There have been a considerable amount of innovations in the last few decades. Many of these innovations improve construction methods, decrease cost and time, increase thermal efficiency, enhance structural integrity, and more thoroughly protect against wind and seismic forces. Modern innovations can be a great investment in your home. It is important to be sure that the materials you choose are acceptable in your jurisdiction with regard to the Building Code. You will need to have them properly evaluated by the local building official.

Foundation

Foundation systems have been developed that can decrease costs and employ natural materials. These systems are also geared toward ease of use to allow owner-builders to install it on their own. Some of the innovations currently approved by the IRC include wood foundations, frost-protected shallow foundations, and insulated concrete forms.

Wood Foundation

Permanent wood foundation systems are an option in the Code for conventional construction.

> *Wood foundation walls shall be constructed in accordance with the provisions of Sections R404.2.1 through R404.2.6 and with the details shown in Figures R403.1(2) and R403.1(3).*
>
> *R404.2.1 Identification. All load-bearing lumber shall be identified by the grade mark of a lumber grading or inspection agency which has been approved by an accreditation body that complies with DOC PS 20. In lieu of a grade mark, a certificate of inspection issued by a lumber grading or inspection agency meeting the requirements of this section shall be accepted. Wood structural panels shall conform to DOC PS 1 or DOC PS 2 and shall be identified by a grade mark or certificate of inspection issued by an approved agency.*
>
> *R404.2.2 Stud size. The studs used in foundation walls shall be 2-inch by 6-inch (51 mm by 152 mm) members. When spaced 16 inches (406 mm) on center, a wood species with an Fb value of not less than 1,250 pounds per square inch (8612 kPa) as listed in AF&PA/NDS shall be used. When spaced 12 inches (305 mm) on center, an Fb of not less than 875 psi (6029 kPa) shall be required.*
>
> *R404.2.3 Height of backfill. For wood foundations that are not designed and installed in accordance with AF&PA Report No.7, the height of backfill against a foundation wall shall not exceed 4 feet (1219 mm). When the height of fill is more than 12 inches (305 mm) above the interior grade of a crawl space or floor of a basement, the thickness of the plywood sheathing shall meet the requirements of Table R404.2.3. Even footings can be low impact, made of all natural components. R403.2 states in part: Footings for wood foundations shall be in accordance with Figures R403.1 (2) and R403.1 (3). Gravel shall be washed and well graded. The maximum size stone shall not exceed ¾ inch (19.1 mm). Gravel shall be free from organic, clayey or silty soils. Sand shall be coarse, not smaller than 1/16-inch (1.6 mm) grains and shall be free from organic, clayey or silty soils. Crushed stone shall have a maximum size of ½ inch (12.7 mm).*[6]

[6]*2006 International Residential Code, International Code Council, Inc., Section R404*

Frost-protected Shallow Foundations

Freezing damages a concrete or masonry foundation system through frost heave, which is the expansion of water due to freezing within subterranean soil. Due to the possibility of damage to the foundation, the Code requires protection against freezing. Normally this involves placing the footing below the frost depth of the soil, which varies according to the climate. A frost-protected shallow foundation (FPSF) system is one that does not extend beneath the frost depth but rather is protected in other ways. By using insulation around the foundation, it allows less concrete to be used, less earth to be affected, and still provides an adequate support for bearing walls. FPSFs are permitted to be used for conventional construction. You use less concrete and turn less earth with shallow foundations. The 2006 IRC allows the use of this type of foundation.

> *For buildings where the monthly mean temperature of the building is maintained at a minimum of 64°F (18°C), footings are not required to extend below the frost line when protected from frost by insulation in accordance with Figure R403.3(1) and Table R403.3. Foundations protected from frost in accordance with Figure R403.3(1) and Table R403.3 shall not be used for unheated spaces such as porches, utility rooms, garages and carports, and shall not be attached to basements or crawl spaces that are not maintained at a minimum monthly mean temperature of 64°F (18°C). Materials used below grade for the purpose of insulating footings against frost shall be labeled as complying with ASTM C 578.*[7]

Using this type of foundation conserves the use of overall material and the amount of excavation needed. An FPSF satisfies the condition that would otherwise require a foundation to be installed several feet deep in cold climates. Instead of placing foundations below the frost line, an FPSF raises the frost line to just below the surface using insulation and drainage techniques. This allows the footing to be placed as little as 12 inches below grade. The illustration in Figure 5-10 shows how to install insulation in these shallow foundations.

A frost protected shallow foundation is an approved means of building an adequate foundation while using a minimal amount of concrete. The illustrations show how to install insulation in these shallow foundations.

Insulated Concrete Forms

Insulated concrete forms (ICFs) are a unique way of achieving the solid durability of a reinforced concrete wall system while still maintaining energy efficiency with superior air sealing measures and thermal performance. The ICF principle is using insulating material as a form for a poured concrete wall. The result is a wall with thermal insulation on both sides of a reinforced concrete wall. In addition, the ICF can be assembled quickly with very little experienced labor. This type of wall system is becoming more common and is specified in the IRC with prescriptive standards for ICF construction. "Insulating Concrete Form (IFC) walls shall be designed and constructed in accordance with the provisions of this section or in accordance with the provisions of ACI 318."[8]

Masonry and Concrete Wall Systems

Concrete and masonry offer durability and longevity as their contribution toward a green home. In addition, several innovations now make concrete and masonry less harmful on

[7] *2006 International Residential Code, International Code Council, Inc., Section R403.3*
[8] *2006 International Residential Code, International Code Council, Inc., IRC, Section R611.1*

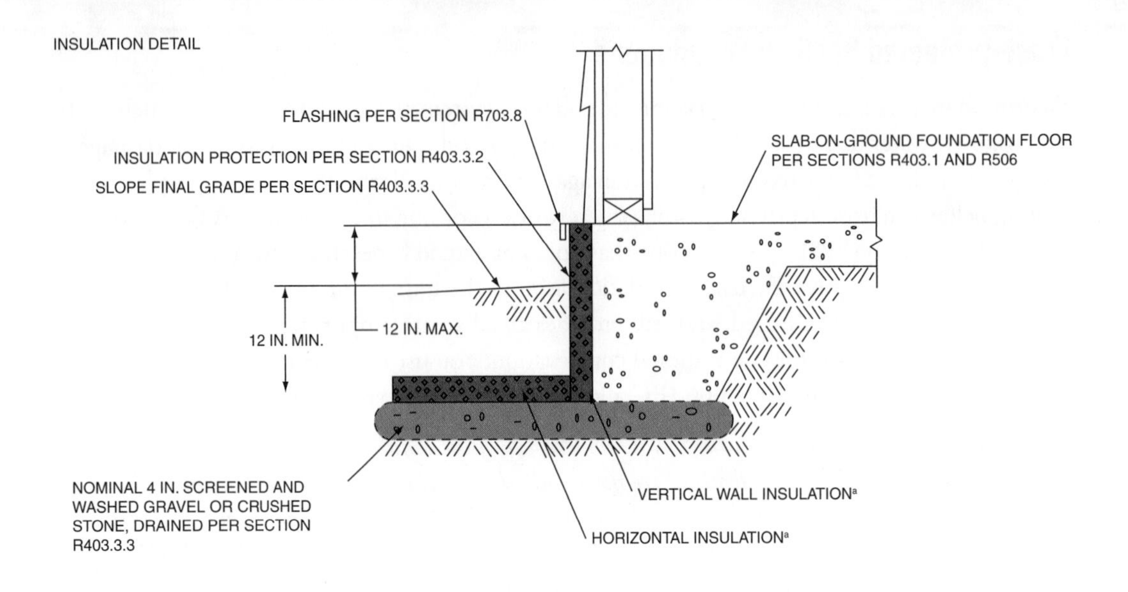

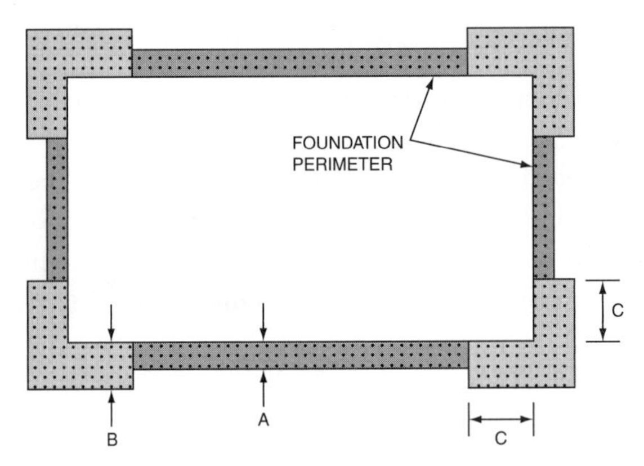

Source: Reproduced with permission from the International Residential Code 2006. Country Club Hills, IL: International Code Council, 2006, Figure R403.3. (1)

Figure 5-10 ■ To reduce the amount of concrete in a foundation, in warmer climates a frost protected shallow foundation does not need to extend below the first line if provided with insulation as illustrated above.

the environment. Precast aerated autoclaved masonry and mortarless block are two modern innovations that are good choices for keeping with your green building plan.

Precast Aerated Autoclaved Masonry

Precast aerated autoclaved masonry is used in making wall units. It is a precast structural product made of cement, lime, water, and sand with a proprietary expansion agent—aluminum powder—added. Aerated concrete wall systems are a lightweight alternative to conventional masonry construction with all the benefits. After mixing and shaping, the product is then autoclaved using heat and pressure. This process allows the expansion to create the properties of insulation and sound proofing. The result is a product that is fire and pest resistant. It is also economical and environmentally friendly. The aluminum causes expansion in the material, which in turn causes the finished block to expand within a mold—a process that includes millions of microscopic hydrogen bubbles to form within the material. This expansion causes decreased density. The material is then cut to size and formed by steam curing in a specialized container (autoclave). The formed block weighs about one-fifth the normal weight of concrete with corresponding thermal- and fire-resistance properties. The expanded lightweight concrete has thermal and acoustic insulating values. Autoclaved aerated concrete (AAC) has significant thermal insulating qualities. According to industry representatives, a low thermal conductivity (U value) combined with the thermal mass effect results in a 10-inch wall that yields an R-31 equivalent rating although the Autoclaved Aerated Concrete Products Association (AACPA) indicates that the effective insulation R value depends on the geographic location.

An option to improve the structural integrity of autoclave masonry is to reinforce the AAC panels with steel or welded wire frames, which are placed into a mold prior to pour. Once the slurry is poured in, the mixture begins to expand around the reinforcement and the composite assembly will act much like reinforced concrete. An AAC has many advantages, including that it is fire resistant, noncombustible, lightweight, durable, and strong; has excellent insulating properties (about R-3 per inch); is versatile in design (can be produced in many desired shapes); doubles as an acoustic insulation; installs quickly (reducing labor costs); is termite resistant; and is considered a green building material.

Mortarless Block

Mortarless block wall systems are not new but are used successfully in a variety of projects, including residential and commercial buildings. Commonly known as dry stack systems, the process involves eliminating the use of mortar between the joints of blocks stacked to form a wall. The mortar would normally serve to bond the individual units together and resolve irregularities in size and shape by adding or subtracting mortar. The mortar, when installed properly in bed joints and head joints, serves to create a bond that holds the wall system together, sometimes even without added grout. A mortarless wall system relies on other means to securely fasten the block units together. In some assemblies, the wall system is formed with interlocking blocks that connect both horizontally and vertically.

In this case, either a surface bonding technique or solid grouting is necessary to ensure stability. Although more concrete is used in the case of solid grouting, in either case, the assembly time is reduced by avoiding the installation of mortar.

Light Gauge Steel Framing

Light gauge steel is a modern product but is considered mainstream because it has become so prevalent. This is very similar to wood (stud) framing except that steel members are used instead. Members are spaced interstitially within a wall frame that is clad with sheathing: siding or wallboard. Steel framing, as illustrated in Figure 5-11, is very strong. It is 100% recyclable and is inorganic and therefore not subject to irregularities, rot, or warping. The material quality is consistent because the product is manufactured. It is easy to assemble, is fire resistant, and has significant structural integrity.

Light gauge steel framing is a popular alternative to wood framing. This technique uses framing elements similar to wood framing in width and length but that are formed as channels, providing a strength axis. There are 3½-inch- and 5½-inch-wide channels that would substitute for 2 × 4 inch or 2 × 6 inch wood framing.

One disadvantage of steel framing is that of thermal bridging. Heat transfers faster through metal than other materials. Therefore, the heat loss experienced (from either side) is higher in

Courtesy of iStock Photo

Figure 5-11 ■ Light gauge steel framing is a very popular material choice for a green home. With virtually no waste, elements of steel framing are normally connected together with screws.

a steel frame. Without a thermal, insulated break, additional energy loss will be realized. This heat loss is conductive through the steel stud and plates. Metal roof trusses similarly transfer conductive heat loss from conditioned space into attics.

Engineered Wood Products

Engineered wood framing materials represent a significant source where alternative materials are both available and approved for use in your new green home. Some aspects of these materials are engineered and manufactured under controlled processes. Others are designed as trusses or I-joists. All represent advances in proven technology that are more economical. Examples of engineered manufactured floor framing systems include modern manufactured joists made from components or a complete factory-made floor system delivered to the site, complete with decking attached. Dozens of companies have advanced engineering design to improve efficiency, decrease materials, extend performance, and generally use less overall material than conventional wood floor framing. Many innovations are considered green for these reasons. In addition, reduced thermal mass decreases conductive heat transfer and therefore improves energy efficiency.

Structural Insulated Panels

Wall framing systems have undergone a metamorphosis as well. Structural insulated panels (commonly known as SIPs) are manufactured wall systems built in factory conditions, which results in extremely low waste and high efficiency. The panels are commonly built with panel board, such as plywood or oriented strand board sandwiching insulation, built as a frame. These can be designed to support specific span and loading conditions. The SIP has its roots in the 1930s with Frank Lloyd Wright using it in his Usonian design. The first foam-filled core was developed in 1952. SIPs are a modern, green solution for residential construction. SIPs are known as a composite panel building system. Technological advances have allowed the production of SIPs so that a competitive alternative to conventional framing could be offered. The use of SIPs has numerous advantages, including accuracy and truly straight framing. They are made with a variety of sheathing, including oriented strand board (OSB), plywood, or fiber-cement board. They are available for 4-inch and 6-inch walls and are commonly built 8 or 9 feet in height and vary between 4 and 24 feet in width. The longer panels must be hoisted in place with a crane but this is normally arranged with the manufacturer. The insulation is normally polystyrene or polyurethane sandwiched, as illustrated in Figure 5-12. One particular advantage is the air sealing capacity of SIPs. Their thermal performance is significantly better than conventionally framed walls. The flexibility of design, extraordinary structural integrity, along with energy-saving insulation make SIPs a modern building material for green building.

Factory-framed Floors

Factory-framed floors, walls, and roof systems bring manufacturing precision to your job site and all but eliminate waste. These have essentially the same appearance as a field-framed floor system, except that they are produced under factory conditions and are virtually error-free. A specific design is given to a manufacturer and a floor system complete with joists and decking is built, delivered to your job site, and hoisted into place with a crane. Your design will be built exactly to your specifications. Like any innovation, there are advantages and disadvantages that must be weighed. The advantage to using factory-framed flooring is the quality and efficiency realized by factory conditions. The disadvantage is that if your foundation is out of square, your floor system will reflect that error.

Engineered Truss

An engineered truss system, as illustrated in Figure 5-13, is a designed mechanism that relies on comparatively slender members that are connected at the joints to transfer loads to a point of support: normally each end. The truss has been around for a long time and its design has been recently improved. Longer spans are possible and trusses can carry more loads and be

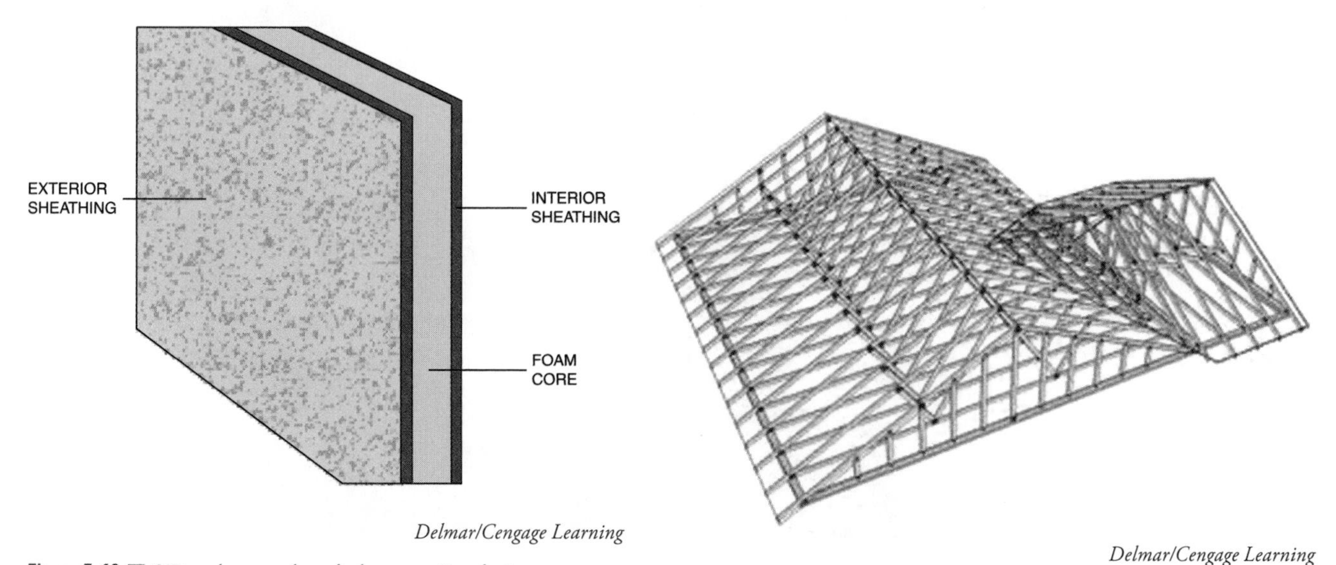

Delmar/Cengage Learning

Figure 5-12 ■ SIPs rely on a bonded connection between exterior and interior structural sheathing to a foam core that provides a thermal barrier from energy loss.

Delmar/Cengage Learning

Figure 5-13 ■ An engineered truss system.

longer than conventional lumber because of connections at the joint. This joint is usually created using a gusset plate. The truss uses significantly less lumber than conventional framing would require. There are different types of trusses identified by shape: common, scissor, Fink, Howe, and many others. Some trusses are designed to avoid wall framing, creating a room or space. These are called studio or room trusses.

I-Joists

I-joists are engineered products that fall somewhere between a solid sawn member and an open web truss. They resemble solid framing lumber but rely on a top and bottom chord to hold the structural element—the web—in place. There are many different manufacturers of I-joists, all of whom have different design criteria for their product. An example of a typical I-joist in Figure 5-14 illustrates how versatile these structural components are. Do not assume that a 14-inch I-joist of one manufacturer is interchangeable with a 14-inch I-joist of another manufacturer. Each may have a different maximum span or center spacing requirement. They may have differing installation methods and even connection details. Some manufacturers can allow for a cantilevered condition. Others will not.

Courtesy of photos.com

Figure 5-14 ■ An I-joist is a generic term for an engineered system that replaces conventional dimensional lumber. An I-joist typically has a greater span while using less nonrenewable resources such as wood timber.

Laminated Lumber

Engineered lumber, such as glulams, structural composite lumber, or finger joint studs, are relatively new innovations but are now almost commonplace on job sites. A glulam is a stress-rated, engineered wood product comprised of wood laminations attached together with glues. The laminations can be from a variety of tree species and are 2 inches or less in thickness. This is a beautiful beam when finished and is often left exposed for appearance. Structural composite lumber includes laminated veneer lumber (LVL), laminated strand lumber (LSL), and oriented strand lumber (OSL). These are all made by stacking wood strips or veneers with adhesive and curing them. LVL, LSL, and OSL are normally available in conventional sizes that match construction practices.

Plywood and Oriented Strand Board (OSB)

Plywood is manufactured using wood veneer that is cross-laminated and sealed with heat, pressure, and an adhesive appropriate to the application. Plywood is the *original* engineered lumber.

OSB is manufactured from wood fibers with heat-cured adhesives. The layers of these fibers are arranged in a cross-oriented pattern, similar to plywood. This makes a rigid wood panel with similar characteristics of plywood. OSB is used in numerous applications, including panel sheathing, I-joists, floor and roof bracing, shear wall brace panels, engineered lumber as laminations, as well as roof and floor decking.

Finger-Jointed Lumber

The finger joint process allows short pieces of lumber to be used as full-length members for studs or plates. The process uses prescribed cutters and adhesives to create a full-height length for a framing member. This method has become a popular way of avoiding waste while making use of short sections to achieve maximum profit. There are reasons that a green home would intentionally use finger-jointed lumber. This method focuses on a material that would otherwise be hauled to the landfill. Finger-jointing lumber for both structural and nonstructural uses is approved for use by the Building Code.

> *Approved end-jointed lumber identified by a grade mark conforming to Section R602.1 may be used interchangeably with solid-sawn members of the same species and grade.*

Finger joint material makes use of smaller pieces of lumber that would normally be discarded.[9]

Roofing Materials

Roofing materials are another avenue to consider alternative materials that support sustainable design and green building. A variety of material products are regarded as sustainable because of their durability and longevity. They will not need to be replaced as often. In addition, the additives in certain roofing materials are less harmful to the environment. Some materials can leach harmful chemicals through rainfall into surrounding terrain. Some materials maintain more heat than others. This could be good in colder climates or winter seasons but a poor choice in hot climates.

[9] *2006 International Residential Code, International Code Council, Inc., Section R602.1.1*

Metal Roofing

Metal roofing has several benefits for use in a green home. The durability of a metal roof avoids the need to reroof every 20 years. A metal roof can last for the life of the home. Metal roofing can withstand a wide variety of weather and atmospheric conditions and still remain durable. Many metal roofs are manufactured from recycled content and the material itself is recyclable. Metal roofing is lighter than many roofing products as well. This adds less weight to the structure, allowing it to endure longer. The lightweight nature of metal sometimes allows it to be installed over existing roofing, saving landfill storage of the existing material. The metal roof can be painted to aid in reflectivity, which will decrease solar heat gain.

Depending on the care exhibited during installation, there can be some problems with metal roofing that call for care when shopping. Metal roofs are susceptible to a number of problems that can lead to leaks, regular maintenance, or repair. Roofs expand and contract with the weather. Metal roofs are made of rigid material fastened to the substrate, normally with nails or screws. Expansion and contraction tend to loosen screws, pull at seams, and cause cracks to form around these and other penetrations. Snow, rainfall, and standing water eventually cause rust or corrosion that could lead to leaks.

Metal roofs also serve as a thermal barrier, retarding solar heat gain, if properly designed and installed. Conditions such as a properly rated heat-resistant roof membrane system and metal roofing could be satisfied by adhering to Energy Star standards for metal roofs.

Membrane Roofing

Membrane roofing includes materials regarded as environmentally friendly, meaning they are durable, energy efficient, and normally made of recycled materials. These roof systems are low maintenance and relatively easy to install. Additionally, the materials themselves are normally recyclable when their useful service life has expired. These include TPO (Thermoplastic Polyolefin), EPDM (Ethylene Propylene Diene Monomer is a synthetic rubber), PVC (Polyvinyl Choride), modified bitumen, and similar products. Membrane roofs have two specific uses that are directly associated with green homes. First, they can serve as a means for rainfall collection from a roof that is normally channeled toward a cistern. Roofs of this nature are intended to harvest rainwater for domestic reuse. Second, they can serve as the underlayment or roofing material for a green roof. Normally, membrane roofing materials cannot be exposed to the sun and need a covering material to act as protection from solar radiation and ballast to hold them in place.

Green Roofs

Green roofs consist of vegetation in a growing medium planted over a roofing material that is typically an impermeable barrier, such as a rubber roof. Green roofs can be installed in many types of buildings, both commercial and residential. They can be an architectural roofing feature designed to lower solar heat gain by providing shade and thorough evaporation. They can reduce rainwater runoff normally sent to storm water drains. A cistern or drainage collection mechanism supplying irrigation systems could be part of such a system. The plants in a green roof reduce airborne pollution while providing a habitat for birds. The green roof provides an additional layer of insulation, retarding heat gain.

One consideration for the structural design of the roof and wall framing is that they must support the loads imposed on them. Adding a method of water collection to a green roof would increase the load, and the roof structure must be strengthened accordingly. A green roof acts as a cistern, holding a small amount of water to maintain the growth on the green roof. This water adds a load to the structure. An engineer may be needed to establish the necessary increase in design strength. A drawback of the green roof is possible maintenance and repair. Even the smallest pinhole leak could potentially damage or destroy cellulose material under the leak. Regular monitoring is needed to prevent continuing damage of this kind.

Exterior Siding

Exterior siding is a common area of a house for the innovative use of nonconventional manufactured materials. The exterior of a building is the first line of defense against the weather. It is also a means of establishing a seal against air leakage. Modern innovations in siding have produced lots of options from which to select a good alternative that fits any need.

Fiber Cement

Fiber-cement composite is a composite building material made of sand, cement, and cellulose. The material is commonly shaped like clapboard siding and is installed in layers. It is also available in panel sheets. It is very durable once installed and is resistant to insects such as termites. The composite can be molded to look like wood. Its value to a green home is its durability and low maintenance.

Vinyl

Vinyl-type materials have been commonly used as siding for many years. In addition to being recyclable, they can be engineered with durability and longevity in mind. They can withstand strong winds if installed properly. Although these durability characteristics help their reputation, there is an environmental downside to vinyl products. Some vinyl building products are recycled, but the demanding extrusion process of vinyl siding means that using recycled vinyl in siding is not common. Also, the overall LCA of building materials with certain chemistries, such as chlorine-based chemicals, are a matter of much debate and concern. PVC production is alleged to have by-products that contain dioxins, a harmful chemical. Using vinyl is probably not the best choice for a green home.

EIFS

Exterior insulated finish system (EIFS) is an alternative to conventional exterior stucco. It is easily and quickly installed and has a built-in insulation component. It must be installed properly and with care. If the installation is done improperly, the most common outcome will be moisture trapped within the assembly, causing deterioration of the structure. Because of the complexity of this installation, the Building Code requires that a special inspection be performed on any installation involving EIFS. A system such as EIFS that has very low tolerance for error can easily suffer from poor or improper installation and lead to failure. If installed correctly, it represents a green alternative to conventional exterior plaster because it uses fewer natural resources and adds an insulation feature and air sealing measure.

Insulation

Many insulation options have been developed in the last few years. Until recently, fiberglass insulation or blown-in cellulose materials were the only two choices. Now there is a variety of synthetic products, such as sprayed polyurethane foam, expanded polystyrene (EPS), polyurethane, rigid board insulation, spray-on products, natural alternative batt type, and even recycled blue jeans, that are available as insulation.

Building Wrap

Moisture causes damage to most building materials, especially wood. A weather-resistive barrier is intended to prevent moisture from causing that damage. The IRC specifies how to install the barrier:

> *R703.2 Water-resistive barrier. One layer of No. 15 asphalt felt, free from holes and breaks, complying with ASTM D 226 for Type 1 felt or other approved water-resistive barrier shall be applied over studs or sheathing of all exterior walls. Such felt or material shall be applied horizontally, with the upper layer lapped over the lower layer not less than 2 inches (51 mm). Where joints occur, felt shall be lapped not less than 6 inches (152 mm). The felt or other approved material shall be continuous to the top of walls and terminated at penetrations and building appendages in a manner to meet the requirements of the exterior wall envelope as described in Section R703.1.*
>
> *Exception: Omission of the water-resistive barrier is permitted in the following situations:*
>
> 1. *In detached accessory buildings.*
> 2. *Under exterior wall finish materials as permitted in Table R703.4.*
> 3. *Under paperbacked stucco lath when the paper backing is an approved weather-resistive sheathing paper.*[10]

House wrap, as illustrated in Figure 5-15, is an alternative material intended to form a barrier that provides a weather seal, protecting framing and insulation within a wall cavity. Installing this barrier is intended to prevent the penetration of wind, moisture, and the associated pollutants that are atmospherically suspended. Other incidental aspects of building wrap include thermal resistance values to the wall and the ability to deaden noises. An important aspect of building wrap is to shield the substrate and framing elements from weather-related elements while allowing any moisture that penetrates to escape. To achieve this goal, there are many different methods available from a variety of manufacturers. The overall goal of reducing moisture penetration is to maintain the structural integrity of the building and prevent mold. The wrap serves as the water management system, carrying moisture down and out along the drainage plan of the wall.

The Federal Emergency Management Agency (FEMA) has set forth standards for house wrap installation methods for coastal conditions:

- Follow manufacturers' instructions.
- Plan the job so that housewrap is applied before windows and doors are installed.

[10]2006 International Residential Code, International Code Council, Inc., Section R703.2

Courtesy of iStock Photo

Figure 5-15 ■ Typical house wrap is installed over substrate and before finish siding is installed.

- Ensure proper lapping, which is the key—the upper layer should always be lapped over the lower layer.
- Weatherboard-lap horizontal joints at least 6 inches.
- Lap vertical joints 6 to 12 inches (depending on potential wind-driven rain conditions).
- Use 1-inch minimum staples or roofing nails spaced 12 to 18 inches on center throughout.
- Tape joints with house wrap tape.
- Allow drainage at the bottom of the siding.
- Extend house wrap over the sill plate and foundation joint.
- Install house wrap such that water will never be allowed to flow to the inside of the wrap.
- Avoid complicated details in the design stage to prevent water intrusion problems.
- When sealant is required:
  - use backing rods as needed,
  - use sealant that is compatible with the climate,
  - use sealant that is compatible with the materials it is being applied to,
  - surfaces should be clean (free of dirt and loose material), and
  - discuss maintenance with the homeowner.[11]

[11] *FEMA, Technical Fact Sheet #22, http://www.fema.gov/*

Some common problems encountered when installing house wrap include:

- Incomplete wrapping. Gable ends are often left unwrapped, leaving a seam at the low end of the gable. This method works to prevent air intrusion, but water that gets past the siding will run down the unwrapped gable end and get behind the house wrap at the seam. Also, it is common for builders to prewrap a wall before standing it. If this is done, the band joist is left unwrapped. Wrap the band joist by inserting a strip 6 to 12 inches underneath the bottom edge of the wall wrap. In addition, outside corners are often missed.
- Improper lapping. This often occurs because the house wrap is thought of as an air retarder alone. When applying the house wrap, keep in mind that it will be used as a vertical drainage plane, just like the siding.
- Improper integration with flashing around doors and windows.
- Relying on caulking or self-sticking tape to address improper lapping.

Sealant can and will deteriorate over time. A lapping mistake corrected with sealant will have a limited time of effectiveness. If the homeowner does not perform the required maintenance, serious water damage could occur when the sealant eventually fails. Therefore, do not rely on sealant or tape to correct lapping errors.[12]

Fenestration

Doors, windows, and trim are areas of concern for energy leakage. Finding modern innovations that curb the leakage certainly denotes an effort toward a green home. Advanced materials that last longer and provide a more thorough seal are available for your new home. Additionally, certain materials have a lower environmental impact or are recycled (and may be recyclable).

A rating criterion that is accepted nationally for windows and doors is the National Fenestration Rating Council (NFRC). The NFRC is a nonprofit organization that administers an independent rating and labeling system for the energy performance of windows, doors, skylights, and attachment products. A typical energy performance label is illustrated in Figure 5-16. When this council evaluates and rates a window or door, it issues a label indicating the certification.

This evaluation and rating carries information about the product that tells you about its expected performance. Criteria from the NFRC include:

> *U-Factor measures how well a product prevents heat from escaping. The rate of heat loss is indicated in terms of the U-factor (U-value) of a window assembly. U-Factor ratings generally fall between 0.20 and 1.20. The insulating value is indicated by the R-value which is the inverse of the U-value. The lower the U-value, the greater a window's resistance to heat flow and the better its insulating value.*

Glazing Ratings

Solar heat gain coefficient (SHGC) measures how well a product blocks heat caused by sunlight. The SHGC is the fraction of incident solar radiation admitted through a window (both directly transmitted and absorbed) and subsequently released inward. SHGC is

[12]*FEMA, Technical Fact Sheet #22, http://www.fema.gov/*

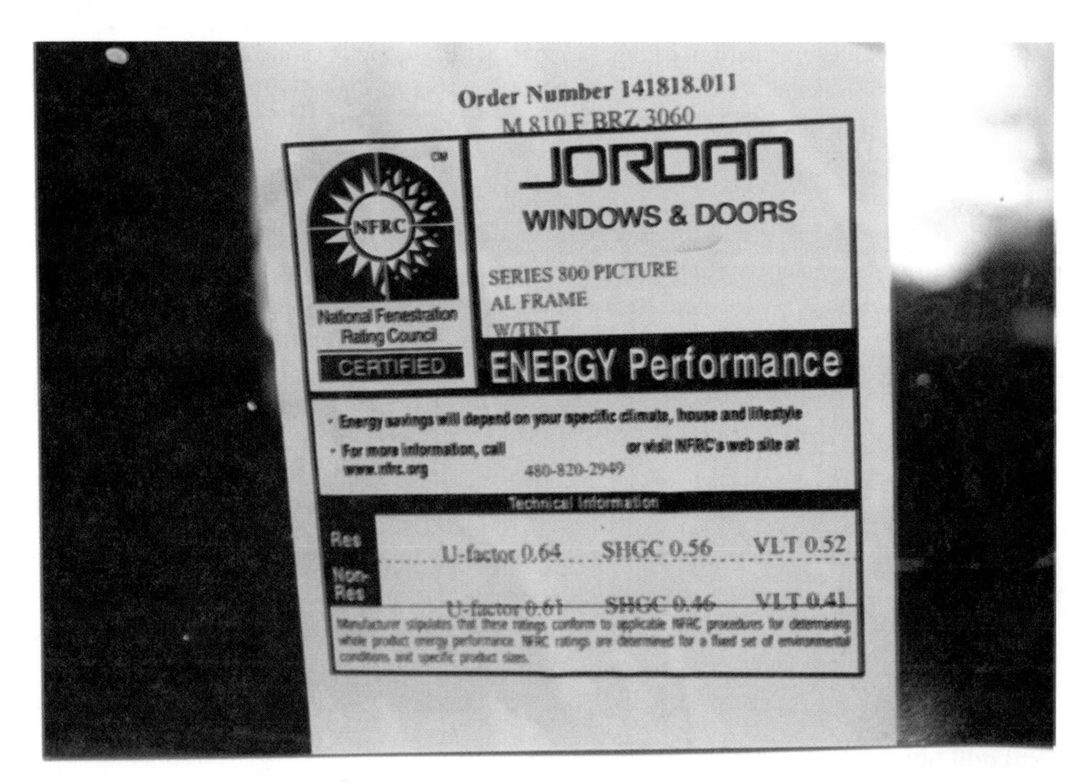

Photo by Lynn Underwood

Figure 5-16 ■ Energy performance rating for fenestration indicates U values and solar heat gain coefficient.

expressed as a number between 0 and 1. The lower a window's SHGC, the less solar heat it transmits in the house.

According to the Department of Energy, visible transmittance (VT) is a measure of how much light comes through a glazing-type product. It is an optical property that indicates the amount of visible light transmitted and is expressed as a number between 0 and 1. The higher the VT, the more visible light is transmitted through the glazing.[13]

Air leakage is indicated by an air leakage rating expressed as the equivalent cubic feet of air passing through a square foot of window area over time (cfm/sq ft). Heat loss and gain occur by infiltration through cracks in the window assembly. The lower the air leakage, the less air will pass through the cracks in a glazing assembly.

According to the NFRC, condensation resistance is a measure of the ability of a glazing product to resist condensation on an interior surface. The higher the rating, the better that product is at resisting condensation formation and is expressed as a number between 0 and 100.[14]

Doors

Alternative door material options include composite or synthetic materials that have either a low impact or boast an extended life and enhanced durability. Additionally, some hollow door

[13] *U.S. Department of Energy, Energy Efficiency and Renewable Energy, http://www1.eere.energy.gov/*
[14] *NFRC, http://www.nfrc.org/*

assemblies have built-in insulation, filling the cavity with thermally efficient filling that retards energy loss and air leakage.

Windows

Shopping for windows that will fit your goal for a green house is easier because they are all graded for thermal efficiency. The efficiency rating is delineated as a μvalue. A μ value is the inverse of an R value, so the lower the μ value, the more thermally efficient the window is. For example, a typical μ value of an efficient window is $\mu = 0.40$ or a U40. This represents an equivalent R value of 1/.4 or an R-2.5. By comparison, for a U60 window, the R value would be 1/.6 = 1.67.

Interior Products

Interior products and finish materials should be a focus for a green building because there are a variety of options to allow for meeting criteria that merit a green home. These interior products include wall and ceiling covering, trim, cabinets, and similar types of materials. There are many alternatives to conventional products that offer improved environmental value.

Wall and Ceiling Covering

Wall and ceiling covering is most commonly achieved with drywall over framing. Because the Code requires an interior finish material to have a limit on the flame spread and smoke-developed ratio, this remains the most common and cheapest option. However, it is essential to select a wallboard with a proper evaluation listing and label to be assured of a safe and green home. Recently wallboard from China seems to have caused certain metals (copper in particular) to discolor and exude a noxious odor. Although an investigation is not yet complete, it is alleged that a higher sulfur content was the cause.

Floor Covering

Floor covering is another area where selection of the proper materials yields some shades of green for your new home. Using a natural material that is low impact or recyclable is a step in the right direction. Although durability is one desirable aspect of materials used in a green home, used floor tile can be recycled to keep it from the landfill. Bamboo flooring is an ecofriendly option that is available for use in a wide variety of applications.

Cabinetry

Cabinetry is another area where you can make a difference. Often, during the lifetime of a home, cabinets are replaced to upgrade its appearance. Anticipating this, a selection of the most modern trend will at least prolong that inevitability. Additionally, the use of composite wood or recyclable synthetic materials to build the cabinets reduces the demand for natural lumber and yields a lower impact on the environment.

Interior Trim

Options for interior trim allow for easy achievement of some shades of green. Synthetic, recyclable material is available for use as trim board. Earth plasters are normally used to cover straw bale and adobe walls and serve as a finished surface. They are made from clay and sand found in the soil. Recycled products or renewable materials, such as bamboo, also make excellent material choices for interior trim.

Plumbing

Plumbing materials include fixtures such as bathtubs, showers, sinks, lavatories, as well as pipes for drain and water supply. The use of water conserving fixtures is covered more thoroughly in Chapter Ten. However, the selection of the right type of plumbing materials is regulated by the IRC and deviations are rare because sanitary safety is necessary. Reuse of currently installed fixtures is one way to be greener. Some limited plumbing materials and fixtures can be reused and thus recycled into newer homes. Ordinarily, pipe is not reusable except under certain circumstances. For example, you may be able to salvage vent pipe. The goal would be to use pipe that is as sanitary as possible. Although the Codes (IRC and IPC) are silent regarding reuse of this kind, they do state that the materials need to be inspected by a third party. This means that certain older pipe may not be acceptable if it does not have that product evaluation and approval. However, common things that may be acceptable to reuse include fixtures such as lavatories, toilets, bathtubs, sinks, and others. There is a drawback if the fixture is the type that is not designed for low-flow requirements. There are some things to consider when shopping for environmentally sensitive plumbing materials.

Green plumbing fixtures are those that will reduce potable water use. Alternately, use low-flow and ultra-low-flow appliances. Most low-flow faucets now limit flow to 1.8 gallons per minute. If you will irrigate a garden, consider drip hoses or sprinkler systems that are programmed to turn on or off. At the very least, you can retrofit sink and lavatory faucets and showers with water-restricting aerators. These restrict the flow rate of water, even under pressure. Normally, potable water is used to propel human waste down the drain. Alternatives to this include low-flow toilets and fixtures. The amount of water is limited to a reasonable amount or a limited flow rate. There are innovations such as a dual-flush toilet with two levers: one for urine and one for solid waste. These use a differing amount of water for the associated flush. Newer systems for urinal use include a nonwater urinal. A urinal of this type uses a chemical sealant that prevents odors and still allows drainage for urine. This sealant, with a specific density, floats on top of the trapped urine, forming an airtight barrier. Urine passes through the sealant, displacing the waste already there. The sealant then allows the effluent to evacuate the urinal. There is no flushing because water is not the method of creating the drainage.

Adding an interior liner to a water or sewer pipe system reduces infiltration from debris, increases the sustainability of the building, and improves water quality. Using an advanced mixture of polymers, epoxy adheres and bonds to the interior of a pipe to create a smooth surface, preventing contamination from developing. This type of treatment prevents infiltration and exfiltration, restores structural integrity, eliminates joints that can weaken, and allows root intrusion. Although this is normally a repair for older, existing pipe that is damaged, installing nontoxic epoxy actually increases flow capacity because it is much smoother than older pipe.

Rain harvesting and storm water collection for reuse in the building's nonpotable water fixtures, such as toilets or irrigation, is certainly a green plumbing principle. Gray water plumbing systems allow used potable water to be reused along the same lines. Although it deals with heating water, on-demand hot water systems can reduce water use waste while you wait for hot water to reach the fixture.

Heating, Ventilation, and Air Conditioning (HVAC)

Selecting efficient HVAC equipment is a significant decision for a green home. The primary purpose of a house is shelter from the weather (cold and heat). HVAC equipment represents a significant part of the overall energy bill for the lifetime of your home, normally around 40% to 60%. Finding the most efficient and economical equipment is critical to having a truly green home. There are ratings available for most equipment.

One such rating is the Energy Star certification program. Sponsored by the federal government's Department of Energy, an Energy Star certification indicates that the appliance has met the required energy-saving criteria.

Radiant heat

Radiant heating is a system used to heat and cool buildings and homes that delivers most of its heat by conduction and convection rather than radiation. Radiant heat transfer is dependent on temperature differential in large part, and most radiant systems operate at very low temperature differentials. Heated water is circulated through floor or ceiling panels to condition the space. Radiant heating has been used for centuries. The Romans used it as a heating method. Architect Frank Lloyd Wright used radiant heat in many of his designs. It is a very cost-effective means of heating a home and it is said to be more comfortable.

DURABILITY AND COST

Durability of a building relates to its environmental impact because it does not make added demands on the environment. A building designed and built to last for 100 years instead of 50 years cuts the environmental impact in half because a replacement of that house is unnecessary. Additionally, the decreased maintenance of the 100-year building carries a corresponding decreased demand on the environment. In most cases, a well-built building also has better thermal performance than one poorly built.

Improved Durability

Improving durability while avoiding environmental impact is truly the goal of a green home. The term *durability* has overarching definitions. Improved durability can take on several aspects, including that of material selection that will last longer or need less maintenance. There are some durability factors that can be managed through design and construction. Adequate control of moisture durability is an issue of water management. The ability of a building to shed water or prevent moisture from collecting simultaneously prevents corrosion of the building elements. Mold springs from mismanaged moisture control. Heat from the sun also causes materials to decay. Heat causes materials to expand and contract. Any dissimilar material can have differential expansion and contraction rates. This causes abrasion and the associated breaking down of the joint or connection. Attic heat raises the temperature of shingles through conductive heat transfer to roof decking and could cause roofing to delaminate or otherwise deteriorate. Managed temperature in all areas improves the life of the house. A deciduous shade tree that shields heat in the summer and allows solar heat gain in the winter is an example of this design. Sunlight in various wavelengths degrades many materials, including plastics, wood, fabric, and paint. Along with heat, this is a major cause of the degradation of roofing material.

Synthetic materials, such as plastics and vinyl siding can be protected against harmful UV rays with treatment. Insects are also responsible for significant damages to buildings. The most notable insect that causes damage is the termite. The Code deals with termite infestation by specifying either treated wood or lumber that is naturally termite resistant such as foundation redwood. Entropy happens and building materials and components deteriorate. Natural forces such as gravity, wind, weathering, or even general use will also cause a material to degrade and become no longer functional. A building that works well responding to weather and everyday life will be durable and have less maintenance over its lifetime. It is essential for a building to be built to withstand any natural event over the expected or desired lifetime in order to be considered durable. They must withstand things like hurricanes, earthquakes, flooding, fires, and similar natural disasters in that region. Although the Code delineates prescriptive methods of meeting these assaults on your home, increased standards that exceed these requirements create more durability.

Cost-Effectiveness

There are many cost issues to consider when building your green home. You need to evaluate the cost of materials and whether purchasing such materials will yield a payback. Repairs, maintenance, and replacement components will be a normal part of your home's upkeep if you do not design and build with durability in mind. There are two broad sources of payback. The first is direct cost. You will pay for it at the beginning or over the long term. Doing it right the first time is far cheaper, easier, and better for the overall building's integrity. For example, imagine that you put poor quality insulation and that was poorly installed. In order to replace this insulation, you would have to take the wallboard completely off and install proper insulation then purchase and reinstall more wallboard. Taking the time and overall cost into account when building your home will allow you to make the smartest and greenest decisions.

BEFORE YOU DECIDE . . . REFLECTIONS AND CONSIDERATIONS

✔ Material selection is a significant aspect of your green home. You should invest in a considerable amount of research before you place your material order and agree on a payment schedule. There are lots of considerations, including:

- Environmental impact of the material selected. This includes embodied energy weighed against the energy saved over the lifetime of the building.
- Cost and value.

- Availability of the product locally to avoid large embodied energy from having it delivered from further away.
- Structural effectiveness based on regional natural forces.
- Appropriateness for the region.
- Reliability of the material or product.
- Durability over decades without significant maintenance.
- Effectiveness in conserving energy.

✔ Conventional materials, such as wood, concrete, or masonry are still found in green homes. Modifying the way a building is built may make it greener.

✔ Used, recycled, or reclaimed materials such as concrete, wood, and masonry are considered to be green.

✔ Natural or renewable materials such as adobe, straw bale, cork, or bamboo are found in green homes.

✔ Modern innovations in foundation, wall, and roof systems as well as fenestration increase energy efficiency or reduce environmental harm.

- Use of innovative materials or systems includes siding, insulation, roofing material, building wrap, wall and ceiling covering, cabinets, and interior trim and plumbing and mechanical systems.

✔ The durability and cost-effectiveness of a material lends itself toward being green.

✔ The reason for material selection should be based on the shade of green that most appropriately fits your commitment to this objective. It is a personal choice and should match your budget.

FOR MORE INFORMATION

Autoclaved Aerated Concrete Products Association
http://www.aacpa.org/

Department of Energy
http://www1.eere.energy.gov/

Environmental Protection Agency *Environmental Building News*
http://www.buildinggreen.com/

EPA Life Cycle Analysis
http://www.epa.gov/

FEMA
http://www.fema.gov/

Forest Stewardship Council
http://www.fscus.org/about_us/

Green Building Press
http://www.greenbuildingpress.com/

National Fenestration Rating Council (NFRC)
http://www.nfrc.org/

Chapter 6

INDOOR AIR QUALITY

INDOOR ENVIRONMENT

Indoor air quality is the measure of the cleanliness of the interior environment because it affects the air we breathe. Although there are definable measures for safe thresholds of poor indoor air quality, there are several things that mitigate these conditions. It is impossible to predict every living condition that could occur in every part of the country. Become aware of the pollutants that are brought into the house and the effect they will have on your home and your health. Start with the design and continue with the construction process and possibly the remodeling process. Chemicals, pets, and other allergens also play a role in indoor air quality. Maintain all equipment such as heating and cooling units, water heaters, ranges, ovens, and clothes washer and dryer in optimum operating condition and avoid long-term moisture conditions. A green home includes improved levels of indoor air quality.

HEALTHY HOUSES

Your green home will be a healthy place to live in if you observe a few simple considerations for indoor air quality. There are many potential sources of pollution within a home. Some enter with outdoor air and others are brought in with furniture, food, clothing, gasoline, soap, and cleaning supplies. Additionally, some pollutants are produced from permanent fixtures in the home that are installed during the construction process such as treated wood, paint, varnish, and similar chemicals. Elements of these sources inadvertently mix with the indoor air we breathe. Indoor air in a home can be seriously polluted and cause health-related issues. Environmental Protection Agency (EPA) studies of human exposure to air pollutants indicate that indoor air levels of many pollutants may be 2 to 5 times higher than outdoor levels and, in some cases, 100 times higher. This is significant because of the amount of time people spend inside their home. Indoor airborne pollutants that have been increasingly recognized as threats to people's health include products of combustion, chemicals introduced from products brought into the home, pets, and those health hazards from people.

Indoor Air Quality Factors

Factors that affect indoor air quality are legion. They include everything from dust to mold. Even human skin particles that become airborne affect the quality of the air we breathe. We ingest some of these skin particles, and the rest are a food source for the dust mite such as that illustrated in Figure 6-1.

About 40 pounds of dust accumulate in an average size house every year. Dust carries dust mites and other illness-causing bacteria and viruses. Pollen, ragweed, and a variety of other allergens find their way into the home from outdoors. Carpet, plywood, and other construction materials emit gas chemicals used to manufacture these products. Mattresses and clothes treated with fire retardants and other chemicals are vaporized and mix with the air we breathe. Heating, ventilation, and air-conditioning (HVAC) air ducts serve as a fertile breeding ground for mold spores. Household cleaners, aerosols, insecticides, pesticides, paints, solvents, chemical fumes, and toxins are a few sources of pollution in our interior atmosphere.

Mold spores, bacteria, and mildew thrive in damp conditions such as kitchens, bathrooms, laundry rooms, or other humid areas. Mold is readily visible as a discoloration on walls and ceilings, as shown in Figure 6-2. Airborne pet dander is also a common trigger for allergies. Insect and vermin feces become airborne in most homes. Smoking in the home adds toxins to the air. Vehicle exhaust and gasoline fumes easily enter your home when you enter your garage.

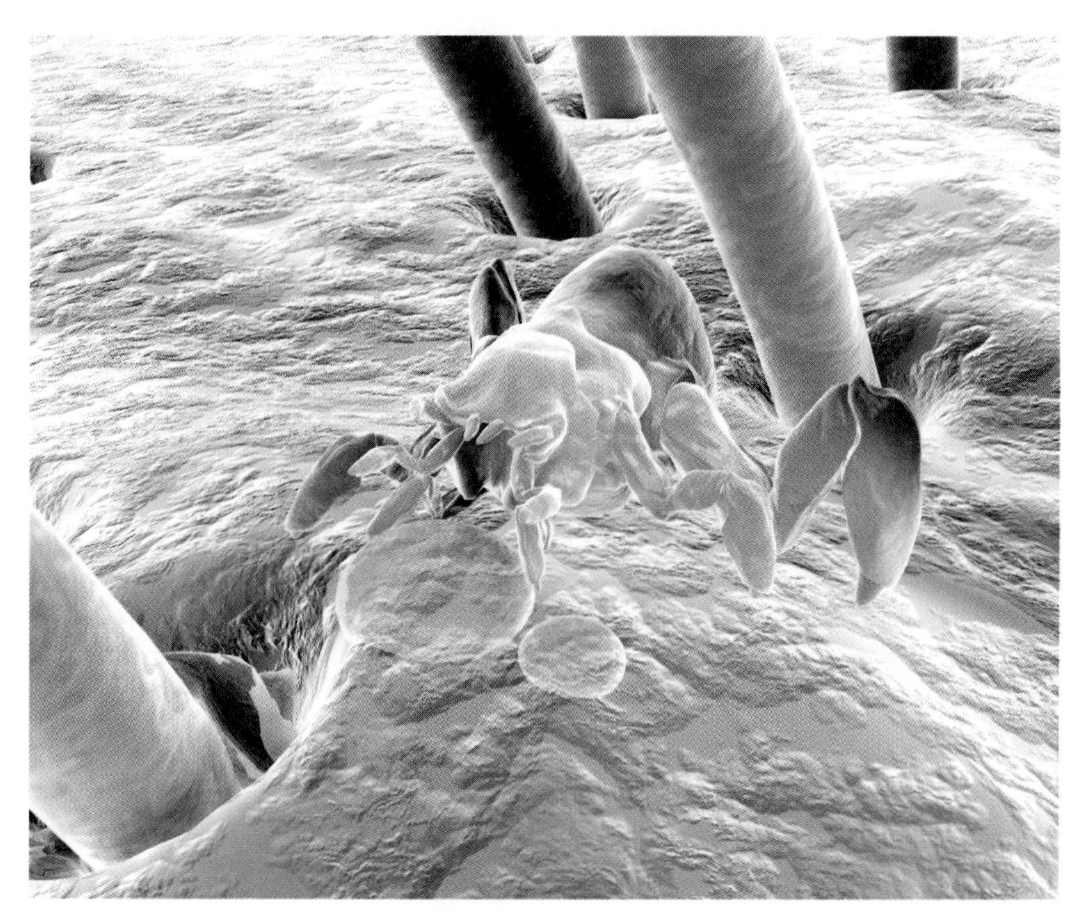

Courtesy of iStock Photo

Figure 6-1 ■ Dust mites are present in insanitary conditions and bring disease to every part of your home. Good indoor air quality includes proper control of these and similar vermin.

Courtesy of iStock Photo

Figure 6-2 ■ Mold represents a health hazard to humans. Mold growth is supported by insanitary conditions, including damp and wet conditions.

Some of the elements that affect indoor air quality are the result of lifestyle choices like smoking. Others are a result of the construction process such as material selection or site selection. Still others are the result of the condition and operation of equipment or building maintenance and cleaning. Each of these factors is within your control.

Biological Factors

Biological factors include things like volatile organic compounds, radon, formaldehyde, second-hand smoke, pet hair, pollen, bacteria, and viruses. The National Academy of Sciences/Institute of Medicine issued a report on asthma and indoor air quality, confirming that dust mites and other allergens, microorganisms, and some chemicals found indoors are triggers for asthma. Biological factors cause many allergic reactions and worsen asthma. These small organic elements are found throughout our home in beds, carpet, and any place where they can flourish. These are thought to be a source of serious, potentially life-threatening diseases such as legionnella, a bacteria species that is thought to cause Legionnaire's disease. But other biological factors are more common. Common insects such as roaches, rats, mice, and mosquitoes are known to carry disease. Components of HVAC systems can be a haven for biological agents.[1]

An HVAC system can easily collect biological agents by siphoning them into the system from the air handler where they grow and flourish then are blown back in throughout the home with conditioned (heated or cooled) supply air. Dust and debris may be deposited in the ductwork or distribution boxes of the air handler. Those biological agents that remain trapped

[1] *The National Academy of Sciences/Institute of Medicine (2000), Clearing the Air: Asthma and Indoor Air Exposures, Washington, DC: The National Academy Press.*

in the system replicate and grow then spread throughout the home, creating the sick building syndrome.

Lifestyle

Lifestyle choices affect indoor air quality. Secondhand smoke is suggested to be a cause of illness. According to the EPA, secondhand smoke is the third leading cause of lung cancer and is responsible for an estimated 3,000 lung cancer deaths every year.[2] About 1,000 of these deaths are people who never smoked. Exposure to secondhand smoke can have serious consequences for children's health, including asthma attacks and respiratory tract infections (bronchitis, pneumonia), and it may cause ear infections. In addition, the EPA estimates that asthma afflicts about 20 million Americans, including 6.3 million children. Since 1980, the biggest growth in asthma cases has been in children under 5 years old. In 2000, there were nearly 2 million emergency room visits and nearly half a million hospitalizations due to asthma at a cost of almost $2 billion and causing 14 million schooldays missed each year.[3]

Of course there are more common household practices that directly affect indoor air quality. For example, using and storing excessive household chemicals increase the potential for a polluted interior environment. Pesticides, cleaning products, solvents, and a hundred other things contribute to poor interior air quality. Other lifestyle choices that adversely affect our indoor air quality are things like cleanliness, food preparation, and pet maintenance.

Equipment Maintenance

A maintenance schedule for all mechanical systems in the house is essential. Knowing how often the HVAC filter is changed and how often equipment is serviced helps to avoid undesirable conditions. Set up a schedule for all required maintenance for every moving part in your home, and then adhere to that plan.

Proper building and equipment maintenance is essential to indoor air quality. All equipment is subject to age and deterioration. Filters that become clogged become useless as cleaning agents for air, and when they impede necessary airflow to vital systems, mechanical equipment breaks down under the strain. A good building maintenance program begins with the equipment manufacturer's instructions. The manufacturer knows how to get the most useful life out of its product. Follow the maintenance schedule religiously by scheduling maintenance just like you would do for your vehicle. A well-maintained HVAC system works more efficiently and will always last longer. Every aspect of a green building must be maintained, including windows, doors, and their associated screens. These all trap contaminates that can further pollute the interior atmosphere.

Healthy Housing

Healthy housing has some common ground with green homes. How exactly can green building principles be expected to improve public health? The National Center for Healthy Housing worked with Enterprise Community Partners to develop key health-based specifications for its Green Communities initiative. In order for a housing development to meet the Green Communities criteria, builders and property managers comply with certain housing-based health improvements. According to Rebecca Morley, Exective

[2] *U.S. Environmental Protection Agency, http://www.epa.gov/*
[3] *Ibid*

Director at the National Center for Healthy Housing, the following checklist of items are simple steps that, while meeting criteria for a green home, also represent principles of healthy houses:

1. All paints, primers, sealants, and adhesives must meet low volatile organic compound (VOC) levels. VOCs can cause cancer and eye, nose, and throat irritation.
2. All composite wood must not have added urea formaldehyde, which is classified as a substance known to cause cancer by the International Agency for Research on Cancer.
3. All carpets must also meet VOC standards and cannot be installed in areas prone to moisture and mold. This can be expected to reduce asthma, allergies, and other mold-induced illnesses.
4. Proper ventilation is required, which means adding exhaust fans for kitchens, bathrooms, and other areas, not only reducing moisture and mold but also removing combustion products such as carbon monoxide, oxides of nitrogen, and other harmful gases. Fresh air supply is also included, instead of the common practice of simply relying on building leakage to provide the needed air quality.
5. Radon testing and mitigation are required for EPA Zone 1 areas and are highly recommended for EPA Zone 2 areas. Radon is the number one cause of lung cancer among nonsmokers, according to EPA estimates. Overall, radon is the second leading cause of lung cancer. Radon is responsible for about 21,000 lung cancer deaths every year. About 2,900 of these deaths occur among people who have never smoked. On January 13, 2005, Dr. Richard H. Carmona, the U.S. Surgeon General, issued a national health advisory on radon.
6. Reducing other sources of moisture and potential mold is accomplished through use of tankless hot water heaters or drains or catch pans under hot water heaters. Cold water pipes are required to be insulated to prevent condensation, and use of moisture-resistant materials in wet areas also prevents leaks. Ensuring proper drainage from basements and foundation walls and for surface water also contributes to a healthy living environment.
7. Carbon monoxide alarms should be installed in or near areas with combustion sources to help warn occupants of unintended buildup of this potentially fatal gas.
8. Using integrated pest management will not only reduce diseases carried by rats, mice, fleas, and other vectors, it will also reduce exposures to pesticides, some of which are potent neurotoxicants.[4]

EQUIPMENT AND MATERIALS

During the design and construction process, the builder should analyze equipment and materials selected for installation with regard to their impact on air quality. Site selection should also be mindful of air quality. The construction site and building are a common ground for individual components that degrade air quality. Pollutants can have adverse effects on the health of the home occupants. It is important to be aware of unsafe conditions that may develop over time and know how to alleviate these problems.

[4] *National Center for Healthy Housing, http://www.nchh.org/*

Mechanical Equipment

Mechanical systems and fuel-burning appliances are significant to consider for control of indoor air quality. Providing a conditioned environment is basic to the purpose of a home. There are some considerations whenever products of combustion are created inside your home. Combustion safety includes safe use of natural gas, propane, oil, and even wood heating products. Products of combustion are sometimes used to heat or cool your home, heat your water, or dry your clothes. Most commonly, natural gas, liquid petroleum gas, oil, kerosene, or wood is used for this purpose. HVAC systems, water heaters, clothes dryers, cooking ranges, and fireplaces are all common equipment that uses fuel to provide a heating function. However, this equipment can create conditions where dangerous amounts of harmful combustion products are released in the indoor air that you routinely breathe. Combustion by-products such as carbon monoxide, nitrogen dioxide, and sulfur dioxide are some of the dangerous materials that become airborne. Carbon monoxide, which is both colorless and odorless, is particularly dangerous. Carbon monoxide poisonings most commonly occur during the heating season and are generally caused by faulty heating devices or those that are not maintained in proper working order. It is essential to exhaust the products of combustion to the outside any time something is burned indoors. The safety standard for this installation comes from two sources: the manufacturer's installation instructions and the Fuel Gas Code for houses (International Residential Code or the International Fuel Gas Code).

Another approach is to seal any combustion equipment away from habitable spaces. Sealed combustion equipment, fan-assisted furnaces, and water heaters are better when installed within the habitable portions of the home. Combustion equipment uses air in a natural draft to dispose of the products of combustion. Natural draft is the tendency for the warm combustion air to rise up in a chimney. Modern, efficient equipment, such as the gas furnace shown in Figure 6-3, does not waste as much energy or send as much heat up the chimney. It weakens this natural draft that can at times be overcome by conditions that depressurize the house, leading to gaseous leaks and other problems associated with combustion products. A sealed combustion equipment closet separates the gas-fired equipment from habitable spaces and improves indoor air quality.

Space Heating Equipment

Natural draft space heating equipment or water heating equipment should not be installed in conditioned spaces unless it is in a sealed room and has an outside air source. This prevents the fuel burning appliance from using conditioned air for combustion, wasting the energy to accommodate combustion. Air handling equipment or return ducts should not be located in a garage unless they are isolated inside a sealed room and have a source of outside air. This protects the conditioned air from inadvertently mixing with unconditioned, polluted air in the parking garage.

Heating Appliance

A fireplace or a fuel burning appliance (except a cooking appliance) should not be located within a conditioned space unless it is within a sealed room or space. Adequate combustion air from an outside source should be available. Additionally, exhaust gases from these appliances should be vented outdoors. All of these conditions are for the purpose of preserving energy loss and exhausting contaminated gases to the outside. Fireplaces and wood burning and pellet stoves must meet the respective energy efficiency and emission standards for the same reasons.

Courtesy of iStock Photo

Figure 6-3 ■ Properly operating mechanical equipment is critical to improved indoor air quality. A poorly maintained system can allow contaminates such as dust or exhaust to enter the habitable spaces.

Direct Vent Appliance

A furnace or fireplace, whether it is a direct vent, sealed combustion, or isolated in a contained closet, will improve indoor air quality. Direct vent or sealed combustion furnaces must be installed and vented per manufacturer specifications and draw combustion air from outside. If a furnace is to be isolated in a contained closet or room, the area must be sealed virtually airtight, including between the bottom plate and subfloor. Sealed sheathing should cover the walls and ceiling, and a solid access door with a threshold and weather stripping or a gasket seal should be present.

Water Heating Equipment

If a fuel-fired water heater is within a room, it should be either within a combustion closet or otherwise isolated from the conditioned area so that no combustion air is drawn into the appliance. Another approach would be to install it in a garage or other nonhabitable room. However, the necessary volume of combustion air must be available to allow it to function properly. Be sure and check the International Residential Code (IRC) for the required volume of air needed to satisfy combustion demands for particular equipment.

A sealed combustion closet for a water heater located within the confines of a habitable space makes perfect sense when trying to avoid the problems associated with gas-fired equipment. As mentioned earlier, for gas-fired equipment, sealing from habitable spaces improves indoor air quality while still providing adequate safety.

Attached Garage

Garages attached to a home should be thermally separated and walls and ceilings should be insulated and serve as an air barrier. Doors between a conditioned space and the garage should be tightly sealed with a gasket to ensure air sealing. Walls should be air sealed with a gasket under the bottom plate as well as on top of the wall separating the garage from the habitable space. This separates the unconditioned garage from the conditioned home.

Garage Exhaust Fan

A garage contains many toxic substances, including paint, solvent, insecticide, vehicle emissions that contain carbon monoxides, sulfur dioxide, nitrogen oxides and particulates, dust, dirt, soot, and smoke. An attached garage or storage room that stores these common household chemicals, as illustrated in Figure 6-4, or a vehicle should have a fan that exhausts these vapors away from the garage and home. This is important because atmosphere from these spaces invades the conditioned air whenever the door opens.

The EPA recommends either a continuous operation exhaust fan that is capable of moving 50 cfm or an intermittent exhaust fan capable of 100 cfm. The fan should be automatically activated when a door is used and operated for a brief period. An alternative would be an operable window, but this would require regular opening and closing and it would have to be timed to activate upon introduction of the pollutant source (automobile).[5]

Carbon Monoxide

Carbon monoxide (CO) is a toxic gas. Because CO fumes cannot be detected, they can kill before one is aware they are in the home. CO develops in a home from several sources that are all combustion related. Appliances using fossil fuel such as gas stoves, water heaters, room heaters, and furnaces all can be sources for CO. Even letting a car idle in a garage could be a source. Even at lower levels of exposure, this deadly gas causes mild effects that can be mistaken for the flu. These symptoms include headaches, dizziness, disorientation, nausea, and fatigue. Because it is so easy to acquire and install, a CO detector is a simple way to monitor indoor air quality to ensure adequate warning if its presence is detected. If you have a small, single-story home, one CO detector may be adequate. However, larger, multilevel homes should have detectors on each level that are interconnected, much like smoke detectors. There are some units that detect and alarm for either smoke or CO. The CO alarms, newly added as a requirement in the 2009 edition of the IRC, will become more commonplace in the future.

Moisture Control

Moisture control is essential for a healthy, green home. Unabated moisture effects include mold and other aspects of nature that lead to human disease. These same conditions retard

[5]*U.S. Environmental Protection Agency, http://www.epa.gov/*

Courtesy of iStock Photo

Figure 6-4 ■ Normal household paints, fuel, or solvents become airborne and degrade indoor air quality.

the effectiveness of other green features in the home, including indoor air quality, energy efficiency, mechanical equipment efficiency, and expected life of equipment and materials. Moisture damages building materials, which must be replaced. Figure 6-5 shows an example where water has penetrated a wall frame and damaged wallboard. In addition, there are aspects of construction where moisture can be managed more effectively, including the crawl space, subfloor or substrate, concrete placement (including shower), drywall texture, water supply systems, painting, HVAC, surface water control, and rain drains. Moisture that enters a wall cavity that is insulated compresses the insulation, lowering its rating. Moisture that remains can damage or destroy the framing elements and wall substrate.

Rainfall Drainage

Guarding against moisture in your home begins with site planning. A drainable site and a proper foundation are essential to prevent moisture from accumulating and remaining within the foundation and crawl space. Because there are a variety of foundation types, this is accomplished in several ways.

Courtesy of iStock Photo

Figure 6-5 ■ Water from a roof leak has seeped into a wall, causing the paint to buckle and peel.

Diverting rainwater

The foundation should be elevated at least 8 inches above an adjacent grade except for spaces designed as basements. Grade should drop away from the building at a rate of at least 6 inches in the first 10 feet. The slope should either continue or the landscape should be terraced to allow the water to be diverted farther away from the home.

Gutters should be installed on buildings to collect and divert all rainwater. Water from gutters and downspouts should be directed at least 5 feet away from the building. An added feature to connect rainwater collectors would be a rain barrel or a cistern.

Foundation drainage

A foundation drain should be installed around the perimeter of the foundation flush with the bottom edge of the footing. This could be a perforated pipe that can collect and distribute rainwater downhill away from the foundation. The pipe should be covered with some sort of silt protection and gravel base to prevent the perforated pipe from becoming clogged.

The IRC specifies a minimum standard for the drain installation.

> *Drains shall be provided around all concrete or masonry foundations that retain earth and enclose habitable or usable spaces located below grade. Drainage tiles, gravel or crushed stone drains, perforated pipe or other approved systems or materials shall be installed at or below the area to be protected and shall discharge by gravity or mechanical means into an approved drainage system. Gravel or crushed stone drains shall extend at least 1 foot (305 mm) beyond the outside edge of the footing and 6 inches (152 mm) above the top of the footing and be covered with an approved filter membrane material. The top of open joints of drain tiles shall be protected with strips of*

building paper, and the drainage tiles or perforated pipe shall be placed on a minimum of 2 inches (51 mm) of washed gravel or crushed rock at least one sieve size larger than the tile joint opening or perforation and covered with not less than 6 inches (152 mm) of the same material.[6]

High water table

If you cannot drain rainwater by using gravity or you are in an area with a high water table or a basement in your building, a drainage system should be installed to remove water from the footing and foundation. For basements, a subgrade perimeter drain pipe installed that drains to daylight might be necessary. For systems that cannot drain to daylight, a pump may be necessary to evacuate the water collected around the perimeter of the foundation. A course of aggregate surrounding the pipe is preferred to create a capillary break and to mitigate accumulation of radon under a slab. Additionally, a 6 mil polyethylene or approved vapor retarder with joints lapped not less than 6 inches should be placed between the concrete floor slab and the base course or the prepared subgrade where no base course exists. This prevents moisture from migrating under a slab.

Moisture barrier

No matter what type of foundation is being considered, a barrier must be installed to prevent moisture from wicking through. Install a moisture barrier between any construction and foundation and footing that are likely to be damaged by damp or wet conditions. Floors below grade must be damp proofed from the top of the footing to the finished grade. Masonry walls below grade should have not less than 3/8 inch Portland cement parging applied to the exterior of the wall. The parging must be damp proofed with bituminous coating, acrylic modified cement, surface-bonding cement, or similar waterproofing. In areas where a high water table or other severe soil-water conditions are known to exist, exterior foundation walls that retain earth and enclose interior spaces and floors below grade should be waterproofed from the top of the footing to the finished grade. Walls should be waterproofed in accordance with 2-ply hot-mopped felts; 55-pound roll roofing; 6-mil polyvinyl chloride; 6-mil polyethylene; 40-mil polymer-modified asphalt; 60-mil flexible polymer cement; 1/8 inch cement-based, fiber-reinforced waterproof coating; or 60-mil solvent-free liquid-applied synthetic rubber or equivalent.

There are other environmentally friendly options where waterproofing of concrete or block systems must be protected. Thoroseal® is one product that uses an acrylic base to form a barrier against water penetration. Thoroseal® Plaster Mix is a cement-based, polymer modified, waterproof, textured coating for concrete and masonry surfaces. It waterproofs concrete block, cast-in-place or precast concrete, brick, natural stone, and Portland cement scratch and brown. It can be used on interior or exterior surfaces either above or below grade. It will waterproof dams, reservoirs, tunnels, basements, and cisterns and will resist hydrostatic pressure from either the positive side or the negative side.

Crawl space moisture

According to the IRC, the under-floor space between the bottom of the floor joists and the earth under any building must have ventilation openings through foundation walls or exterior walls. The minimum net area of ventilation openings shall not be less than 1 square foot for each 150 square feet of under-floor space area. One such ventilating opening shall be within

[6]*2006 International Residential Code, International Code Council, Inc., Section R405.1*

3 feet of each corner of the building. In colder areas, a mechanical ventilation system provides the necessary airflow.[7]

However, the 2006 edition of the IRC offers an option for otherwise required ventilation that is beneficial to a green home.

Unvented crawl space.
Ventilation openings in under-floor spaces specified in Sections R408.1 and R408.2 shall not be required where:

1. *Exposed earth is covered with a continuous vapor retarder. Joints of the vapor retarder shall overlap by 6 inches (152 mm) and shall be sealed or taped. The edges of the vapor retarder shall extend at least 6 inches (152 mm) up the stem wall and shall be attached and sealed to the stem wall; and*
2. *One of the following is provided for the under-floor space:*
   - 2.1. *Continuously operated mechanical exhaust ventilation at a rate equal to 1 cfm (0.47 L/s) for each 50 ft² (4.7 m²) of crawlspace floor area, including an air pathway to the common area (such as a duct or transfer grille), and perimeter walls insulated in accordance with Section N1102.2.8;*
   - 2.2. *Conditioned air supply sized to deliver at a rate equal to 1 cfm (0.47 L/s) for each 50 ft² (4.7 m²) of under-floor area, including a return air pathway to the common area (such as a duct or transfer grille), and perimeter walls insulated in accordance with Section N1102.2.8;*
   - 2.3. *Plenum complying with Section M1601.4, if under-floor space is used as a plenum.*[8]

The notion for a green home is to extend the insulation to the furthest edge of the exterior envelope of the building and avoid unheated ventilated areas like an attic or crawl space. This prevents heat loss associated with the otherwise required foundation (or attic) vents. Closing off the crawl space in this manner has implications related to moisture. Without a method for controlling moisture, damage will occur to the structural elements of the building. Attics and crawl spaces (basements) may become conditioned, thus controlling moisture altogether.

Other Sources of Moisture

There are many other parts of the construction process that introduce water or moisture laden vapors that can settle and remain in the building long enough to be harmful. Learning about them is the first step toward managing these sources of moisture. For example, the following is just an example of conditions where moisture is used in the construction of a home:

- Anywhere concrete is used and must dry
- Installing masonry grout and mortar
- Applying stucco finish
- Applying water-based drywall texture
- Installing or testing plumbing fixtures
- Cleaning with water

These and similar uses of water can introduce moisture into a home and can lead to damage.

[7] *2006 International Residential Code, International Code Council, Inc., Section R408.1*
[8] *2006 International Residential Code, International Code Council, Inc., Section R408.3*

Moisture infiltration and solutions

Concrete, drywall compound, paint, wall or floor tile and carpet mastic, grout, mortar, and wallpaper are just a few aspects of construction that introduce water into a building. The nature of construction assumes that water will evaporate and leave the building, but that is not always the case.

Moisture can accumulate under tiled surfaces. Tile backing materials are installed under tiles to provide a rigid, firm support. The IRC calls for cement, fiber-cement, or gypsum backers to meet the appropriate American Society for Testing and Materials (ASTM) Standard.[9] Doing so will satisfy conditions to prevent moisture accumulation.

Prevent accumulation in porous materials that connect to the ground by installing a capillary break. For example, beneath a concrete slab, add a 4-inch bed of ½-inch diameter aggregate. The gravel bed should be covered with polyethylene or polystyrene sheeting that will be in direct contact with the slab and ensure that joints are lapped.

Moisture and Wood

Wood framing requires careful attention to moisture control in several areas. As you have learned previously, trees are filled with water. Even after drying to a maximum 19% moisture content, if adjacent to a source of moisture, trees can reabsorb and retain moisture, causing deterioration. The following framing steps are areas where careful attention to detail will help you gain control against the enemy of moisture.

1. Seal the band joist. Seal the band joist to retard moisture entry or accumulation. During the framing assembly, install moisture barriers, particularly at areas that will be difficult to get at when framing is complete. For example, the area underneath a cantilever section of floor that extends over a wall or beam can be difficult to access once framing is complete.
2. Prevent condensation on water pipes. Cold water pipes are subject to forming condensation during humid conditions. Installing insulation around the water pipes will resolve this issue. Install water supply pipe in interior walls to avoid leaks caused by the freeze/thaw cycle. This will help avoid condensation and also make any repairs easier if they are necessary.
3. Elevate wood plates. On exterior walls, ensure that wood plates and studs on concrete or masonry foundations are at least 8 inches above the final grade to prevent wicking of water into the wall frame. This involves raising the foundation at least 8 inches above what will be the finish grade so that the wood plate or studs will be well above any possible wicking action of water.
4. Elevate exterior wood siding. On exterior walls, ensure that siding is at least 8 inches above the outside final grade to prevent wicking of water into the wall sheathing. If house wrap is used, ensure that there are no tears and that joints are lapped per manufacturer's specifications.
5. Elevate wood structural members in the crawl space. For crawl spaces, keep beams at least 12 inches from grade and joists at least 18 inches from grade unless they are specially treated to resist decay. In case you want to avoid wood that has been treated to resist decay, natural heartwood grade redwood is acceptable in lieu of this treated lumber for sill plates.

[9] *2006 International Residential Code, International Code Council, Inc., Section R702.4.2, Gypsum Backers*

Roofing and Moisture

Roofing considerations are paramount to controlling moisture accumulation anywhere in the house caused by rain and snow. There are several means to ensure a durable, long-lasting roof. Controlling roof leaks is essential to avoiding water damage to interior building materials. Providing ventilation that allows for the evaporation of moisture is another approach to avoid moisture damage. Proper maintenance and household user functions help to prevent moisture accumulation as well.

Roofing leaks

Avoid roof leaks by installing roofing underlayment, flashing properly so that water flows down and away from any walls or enclosed space. The manufacturer of the roofing material sets the conditions for installation of its product. Follow these instructions precisely.

Attic ventilation

Provide adequate ventilation to avoid buildup of moisture in the attic. Moisture can damage wood and insulation materials. Ensure that there is at least 1 square foot of venting area for every 150 square foot of projected roof area.[10] (Note that this in the theoretical *footprint* seen from the air, directly above the roof and includes the area of any overhangs without regard for the slope of the roof.) The IRC eases off this requirement by half if at least half of the venting area is in the upper portion of the attic area. In effect, you can have 1 square foot for every 300 square feet of projected roof area if ridge vents are used. Other requirements apply for this condition, so be sure to read the Code.

In addition, the IRC now accepts conditioned attic spaces in lieu of attic ventilation.

Unvented conditioned attic assemblies (spaces between the ceiling joists of the top story and the roof rafters) are permitted under the following conditions:

1. *No interior vapor retarders are installed on the ceiling side (attic floor) of the unvented attic assembly.*
2. *An air-impermeable insulation is applied in direct contact to the underside/interior of the structural roof deck. "Air-impermeable" shall be defined by ASTM E 283.*

Exception: In Zones 2B and 3B, insulation is not required to be air impermeable.

3. *In the warm humid locations as defined in Section N1101.2.1:*
   - 3.1. *For asphalt roofing shingles: A 1-perm (5.7 ´ 10-11 kg/s × m^2 × Pa) or less vapor retarder (determined using Procedure B of ASTM E 96) is placed to the exterior of the structural roof deck; that is, just above the roof structural sheathing.*
   - 3.2. *For wood shingles and shakes: a minimum continuous ¼-inch (6 mm) vented air space separates the shingles/shakes and the roofing felt placed over the structural sheathing.*
4. *In Zones 3 through 8 as defined in Section N1101.2, sufficient insulation is installed to maintain the monthly average temperature of the condensing surface above 45°F (7°C). The condensing surface is defined as either the structural roof deck or the interior surface of an air-impermeable insulation applied in direct contact with the underside/interior of the structural roof deck. "Air-impermeable" is quantitatively defined by ASTM E 283.*

[10] *2006 International Residential Code, International Code Council, Inc., Section R806*

For calculation purposes, an interior temperature of 68°F (20°C) is assumed. The exterior temperature is assumed to be the monthly average outside temperature.[11]

Flashing

Use special care for flashing, counter flashing, valley flashing, or step flashing, ensuring no residual moisture can enter. Follow these special techniques to ensure an impermeable seal. The IRC sets out strict standards for flashing.

> *Roof valley flashing shall be of corrosion-resistant metal of the same material as the roof covering or shall comply with the standards in Table R905.10.3(1). The valley flashing shall extend at least 8 inches from the center line each way and shall have a splash diverter rib not less than ¾ inch high at the flow line formed as part of the flashing. Sections of flashing shall have an end lap of not less than 4 inches (102 mm). The metal valley flashing shall have a 36-inch-wide (914 mm) underlayment directly under it consisting of one layer of underlayment running the full length of the valley, in addition to underlayment required for metal roof shingles. In areas where the average daily temperature in January is 25°F or less, the metal valley flashing underlayment shall be solid cemented to the roofing underlayment for roof slopes under seven units vertical in 12 units horizontal or self-adhering polymer modified bitumen sheet.*[12]

Maintain Dry Conditions During Storage and Construction

Maintain a dry material storage and control during the construction process. This will prevent the inadvertent use of damp or wet materials. Arrange for deliveries to be placed where they will not absorb water or moisture. It may be better to let the material supplier notify the delivery driver of any special conditions such as placement to preserve your materials. Additionally, finish framing the house and dry-in with a basecoat of roofing before you introduce or store materials that would retain moisture or be damaged by prolonged exposure to the weather. Purchase lumber that has been kiln dried to maximum 19% moisture content (15% is better). If any wood materials, including framing or sheathing, show signs of mold, replace them.

Mold-Resistant Materials

Use mold-resistant materials in bathroom areas subject to water splash. Areas like the bathtub and shower should use materials that are highly durable and moisture resistant and without paper-facing, such as cement board, fiber-cement board, fiberglass-reinforced board, or cement plaster.

Water Heater Leaks

A water heater can be subject to leaks. For that reason, the IRC insists that if a water heater is in an area subject to damage, a pan is required.

> *Where water heaters or hot water storage tanks are installed in locations where leakage of the tanks or connections will cause damage, the tank or water heater shall be installed in a galvanized steel pan having a minimum thickness of 24 gage (0.4 mm) or other pans for such use. Listed pans shall comply with CSA LC3.*

[11] *2006 International Residential Code, International Code Council, Inc., Section R806.4*
[12] *2006 International Residential Code, International Code Council, Inc., Section R905*

P2801.5.1 Pan size and drain. The pan shall be not less than 1½ inches deep and shall be of sufficient size and shape to receive all dripping or condensate from the tank or water heater. The pan shall be drained by an indirect waste pipe having a minimum diameter of ¾ inch piping for safety pan drains shall be of those pipe materials listed in Table P2904.5.

P2801.5.2 Pan drain termination. The pan drain shall extend full-size and terminate over a suitably located indirect waste receptor or shall extend to the exterior of the building and terminate not less than 6 inches and not more than 24 inches above the adjacent ground surface.[13]

A water heater must have a drain pan and it must drain through an indirect waste (must travel through air before entering the drain) to a waste receptor (drain). This will ensure that if a leak does occur, it will not damage anything in its path.

Ventilation

Ventilation and exhaust must be considered from an oblique perspective because most of the aspects of indoor air quality include bringing ventilation in and exhausting polluted air out. Pollutants affect the air we breathe whether inside or outside. Everyone knows about smog and thermal inversions that prevent the pollutants we create on a daily basis (vehicle emissions, factories, energy production) from rising up through convection into the upper atmosphere and out of our environment. But there are more serious considerations inside our homes and offices. Everything we try to avoid about the outdoor environment is also found within these buildings.

Great consideration has been taken to avoid letting pollution contaminate your indoor air quality, but as always, you cannot control everything. For those pollutants that find their way in, you need to know how to safely get them out.

Bath Fans

Specialized bath fans are available that turn on coincidently with the light switch. There are fans that activate with humidity sensors, activating when the shower starts to remove moisture from the room. These modern exhaust ventilation fans are more energy efficient but still have sufficient power to exhaust the necessary polluted air outside.

A good option is to select a fan with an Energy Star rating. This rating means that the product meets strict energy efficiency guidelines set by the U.S. EPA and the Department of Energy. These fans

. . . include lighting use 70% less energy on average than standard models, saving $120 in electricity costs over the life of the fan. These fans are more than 50% quieter than standard models. They feature high performance motors and improved blade design, providing better performance and longer life.[14]

Ceiling Fans

Ceiling fans are a great way to provide an alternative to cool the air indoors. The act of blowing air instead of removing heat as with a conventional air conditioner is far cheaper. Air blowing

[13] *2006 International Residential Code, International Code Council, Inc., Section R806.4*
[14] *U.S. Environmental Protection Agency, http://www.epa.org/*

on the skin causes evaporation and cools it. In the summer months, you can tolerate a higher thermostat setting and feel the cooler air with a ceiling fan. During the spring and fall, you can use the ceiling fan instead of the air conditioner. These practices will save dollars and energy. In the winter, reversing the direction of the paddle fans will move air around, pushing the hot air down and adding to the comfort level. Overall, the ceiling fan is much more energy efficient for the service provided. Similarly, the use of Energy Star-rated fans means saving energy costs, better performance, and durability.

House Ventilation

Controlled house ventilation is a solution to the stale air that develops in a tight house. Remember, your energy-efficient home has sealed energy leaks in many places. This efficiency could result in lack of fresh air unless resupplied. Air-conditioning or heating systems unless sealed generally are regarded as *closed systems*, because they condition then recycle the existing air. Without exhausting some of the air, your effort to have a wholesome environment will be counterproductive. There are systems that exhaust stale air and allow fresh air into the home. There are also mechanisms to mix the air in both heating and cooling seasons. There are heat recovery ventilator (HRV) systems that use the energy in the stale, polluted indoor air to preheat the fresh air brought in. All reflect the goal to better control the necessary ventilation of the house as a whole.

In most homes, the main ventilation component is an exhaust fan that places the home in a slight negative pressure. This exhaust draws outside air into the home through a variety of sources, including passive air inlets. It can include cracks between material joints, uninsulated shafts, and wall systems. These sources bring in more than just air. They bring dust, mold, insects, and other unwanted pollutants. Exhaust vents are popular mainly because they are generally less expensive. They do help prevent moisture that can cause damage from migrating into the building.

Another approach to improved indoor air quality for a green home is using the guidelines from the American Society of Heating, Refrigerating and Air-Conditioning Engineers.

ANSI/ASHRAE Standard 62.2, *Ventilation and Acceptable Indoor Air Quality in Low-Rise Residential Buildings*, is a new, nationally recognized indoor air quality standard developed solely for residences. This standard covers ventilation for acceptable indoor air quality and serves as a guide for ventilation rates and other measures that improve indoor air quality in homes.[15]

This standard was developed by the American Society of Heating, Refrigerating and Air-Conditioning Engineers, Inc. (ASHRAE). It was approved by the American National Standards Institute (ANSI) as an approved standard for improving indoor air quality. Appliances such as gas ranges, ovens, and gas burning appliances as well as fireplaces and wood stoves rely on a positive pressure to effectively operate. Enough negative pressure would cause flue gases that include harmful chemicals to be drawn into the habitable living space. An option would be to use direct vent appliances or to install sealed combustion appliances that draw air from outside. Another way to achieve a balanced airflow is to use two fans: one to exhaust stale air and another to provide fresh air. A balance in airflow is difficult to achieve. Detailed duct size, blower size, exhaust volume, effective design, and careful installation are necessary. The system must also be somewhat balanced by adjusting flow volume.

[15]*ASHRAE, http://ashrae.org/*

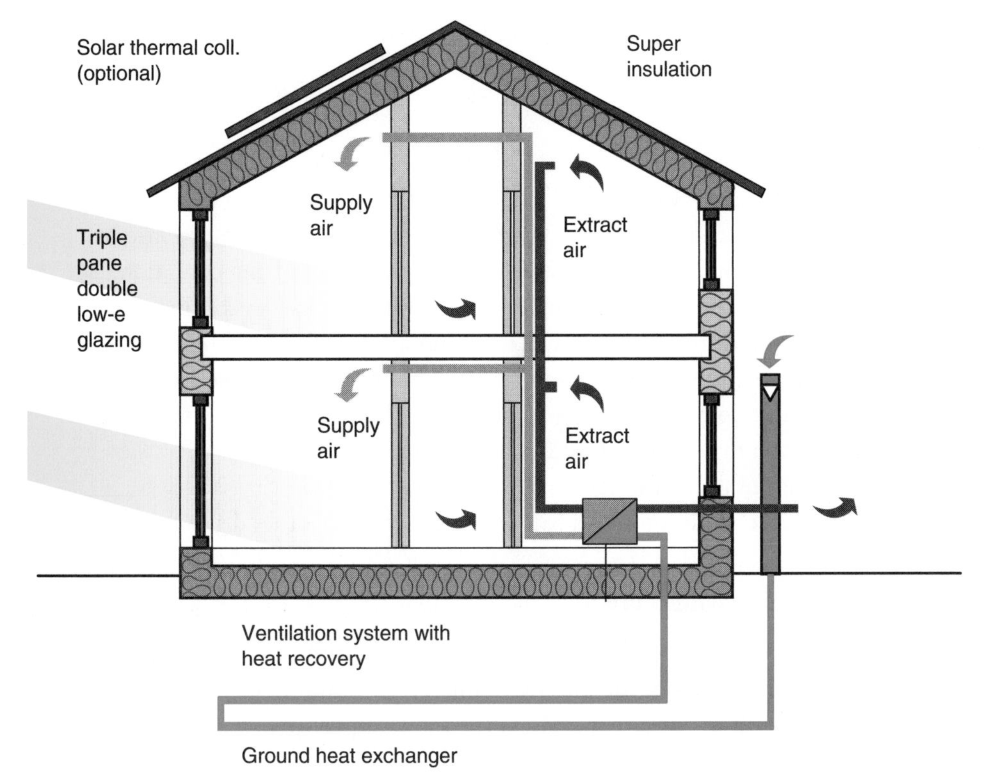

Lynn Underwood

Figure 6-6 ■ Controlled ventilation systems like these can collect the outgoing air into a single duct, making it possible to capture the heat or energy from that air with a heat recovery ventilator (HRV).

Controlled ventilation systems like these can collect the outgoing air into a single duct, making it possible to capture the heat or energy from that air with an HRV. The most common type of HRV is an air-to-air heat exchanger, as illustrated in Figure 6-6. It transfers thermal energy (heat) from the outgoing airstream to the incoming one. As the stale, humid air exits through a heat exchanger, it transfers the heat energy to the air entering the home at the same juncture. In these systems, airflow is balanced. Using exhaust air heating recovers energy, thus effectively reducing consumption.

Blower Door Duct Tightness Testing

To be sure that your home is tight and has no significant leaks you should perform blower door and duct tightness test. These are normally conducted by third-party personnel not hired by the contractor and who report the results to the owner. A standard blower door test has three parts: a calibrated fan that produces pressure, a door panel system, and a tool to measure building pressure. For the test, the blower door fan is sealed. The fan attempts to draw air in by suction or push air out of the building, which creates either a positive or negative pressure difference between the interior and exterior of the building. This pressure difference forces air through any hidden holes or penetrations in the building. The tighter the building, the less air or pressure you need from the blower door fan to create a difference in building pressure. A similar evaluation, the duct blaster test will verify that the home's HVAC system does not have substantial leaks. The blower door test will pressurize the house instead of drawing more air than an acceptable standard (3 pascals). This test evaluates the entire mechanical system, including the exhaust fans, clothes dryer, air handler, and similar equipment. Although atmospheric conditions will affect the results, the test gives you a good idea for how well your system will avoid heat loss through its delivery system.

Range hood exhaust

A kitchen range hood seems essential to rid the house of pollutants such as steam and smoke from cooking. However, if the exhaust is too powerful it can create a negative pressure in the building, sucking beneficial, conditioned air out and creating a back-drafting condition. This condition will siphon air from the outside through the easiest place possible: chimneys, plumbing, or exhaust vents. This would introduce unexpected pollutants inside your home. When shopping for a range hood, be sure to find the appropriate size and manage its use effectively. A 100 cfm fan works best for most domestic cooking appliances. If, for some reason, you think you need a more powerful fan, you may need to install makeup air to balance the air exhausted by the fan. Makeup air is a means of filling the void in a room or space with air that is used for another purpose, such as to support combustion of a fuel source or serve as clothes dryer exhaust.

Clothes dryer exhaust and moisture

Drying washed clothes inside a home is a relatively new phenomenon. A mere 50 years ago, almost all clothes were dried outside on clotheslines. The Building Code caught up with the technology and has built-in rules that address the safety aspect of exhausting moisture-laden vapor from clothes dryers to the outdoors through ducts. There are rules about the maximum length of the duct and how many bends are permitted. Generally, the dryer exhaust, as depicted in Figure 6-7, is limited to 25 feet in length with reductions from this length established for each bend or change in direction of the duct. All these rules are meant to ensure that lint and other debris do not inhibit wet air from getting outside safely. Of course, the dryer manufacturer's instructions prevail if they are tested and listed by an outside agency. It is essential to eliminate the moist air from the house. However, like the kitchen exhaust fan, if excess air is removed, makeup air will come from somewhere, planned or not. If the dryer is in a nonhabitable (and therefore nonheated) space, the impact is minimal. But if within a habitable space, provisions must be made to replenish that air exhausted to prevent negative

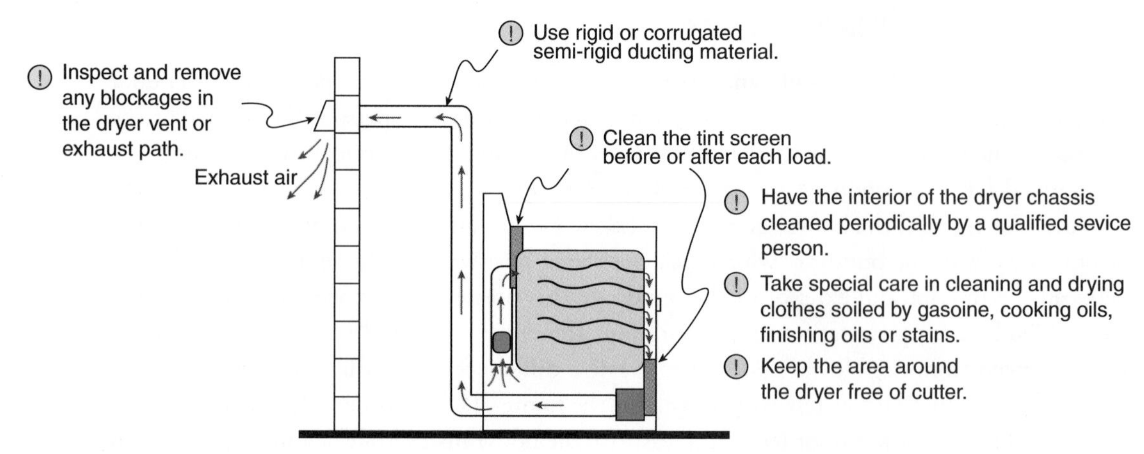

Lynn Underwood

Figure 6-7 ■ Drying clothes exhaust includes moist air as well as lint drawn from clothes. This moist air would otherwise create moist conditions that support mold and other unhealthy conditions.

pressure from drawing the replacement air from an unwelcome source such as concealed wall frames, within stair framing, or chimney chases that contain debris.

Along with the manufacturer's recommendations, the Code has conditions for clothes dryer exhaust as well.

> *The maximum length of a clothes dryer exhaust duct shall not exceed 25 feet (7620 mm) from the dryer location to the wall or roof termination. The maximum length of the duct shall be reduced 2.5 feet (762 mm) for each 45-degree (0.8 rad) bend and 5 feet (1524 mm) for each 90-degree (1.6 rad) bend. The maximum length of the exhaust duct does not include the transition duct.*
>
> *Exceptions:*
>
> 1. *Where the make and model of the clothes dryer to be installed is known and the manufacturer's installation instructions for the dryer are provided to the building official, the maximum length of the exhaust duct, including any transition duct, shall be permitted to be in accordance with the dryer manufacturer's installation instructions.*
> 2. *Where large-radius 45-degree (0.8 rad) and 90-degree (1.6 rad) bends are installed, determination of the equivalent length of clothes dryer exhaust duct for each bend by engineering calculation in accordance with the ASHRAE Fundamentals Handbook shall be permitted.*[16]

Ductwork

Ductwork supplies conditioned air to each habitable room or space from the heating equipment. These ducts are generally one of two types: sheet metal (either rectangular or round and connected by mechanical joints) or flexible, nonmetallic, synthetic material that is factory made to expand and collapse like an accordion. Both are adequate to perform the task, but over time any surface irregularity inside exhaust ductwork such as this can attract debris and mold. To ensure that you always have clean, conditioned air, have your dryer ducts cleaned every so often. If readily accessible, this can be as simple as removing the dryer vent connection and vacuuming the exhaust duct. Cover the ends of the ducts during the construction process to prevent sawdust, smoke, vapors, soil, and other pollutants from entering before you use your heating system. Make sure they are protected until completion of construction. Duct tightness is essential and was discussed elsewhere.

Vacuum System

A central vacuum system, typified by Figure 6-8, is another great way to improve the indoor air quality of your new green home. A central vacuum system consists of a collection unit, piping system and inlet valves, and a hose or wand to collect dust. PVC pipe for transferring soil is installed during the framing stage within the wall, floor, or roof framing and is normally airtight. A hose with power controls attaches to an inlet and activates the cleaning mechanism. Manufacturers have various kinds of equipment and collection options, including bags and filtration. The vacuum system spins, creating the suction and drawing dust-laden air to gather in the canister and be separated by the filtration process. The single most important criterion of this practice is to provide an outside collection receptacle for debris that is away from the living areas. Because the unit is normally installed in a garage, away from habitable spaces, central vacuums ensure that no dust is reintroduced into your home.

[16]*2006 International Residential Code, International Code Council, Inc., Section M1502*

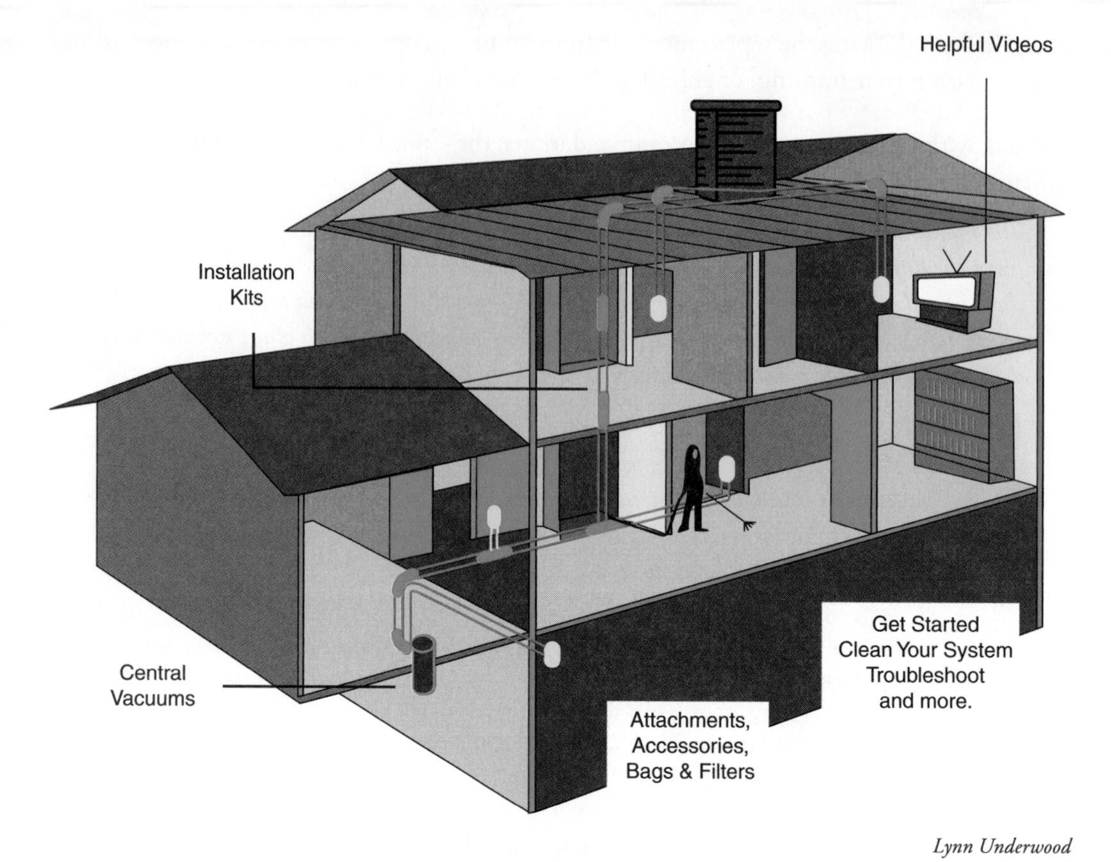

Lynn Underwood

Figure 6-8 ■ Central vacuum systems consist of a collection unit, piping system and inlet valves, and a hose or wand to collect dust.

Volatile Organic Compounds

Building materials such as particleboard, oriented strand board, carpet, paint, wallpapers, cleaning products, glues, and household items such as furniture or cabinets emit volatile organic compounds (VOCs). Even some cosmetics can emit compounds into your indoor environment. VOCs are emitted as gases from solids or liquids. These organic compounds have pressures high enough to vaporize under standard temperature or pressure conditions and enter the atmosphere. Carbon-based organic compounds such as paint thinners, dry-cleaning solvents, petroleum, and gasoline are common VOCs. Other sources include formaldehyde-containing building materials, as well as an array of home and office products ranging from cosmetics, paints, and cleaners to pesticides, copiers and printers, glues and adhesives, and craft supplies. These chemicals give off gas that enters the air we breathe. Some are naturally occurring, such as radon, which comes from within the Earth. However, after entering a home, the gas is trapped and can be ingested. At the extreme end, concentrations of VOCs contribute to creating sick building syndrome.

Formaldehyde

Formaldehyde is one VOC that is common in homes because it is within many building materials. Things like pressed wood, plywood, furniture, insulation, draperies, fabrics, glues, and paints are likely to contain urea-formaldehyde (UF) resins. These materials can be found in subflooring, roof decking, wall sheathing, cabinetry, furniture, and wall covering.

Wood panels as a source

In new construction, the most significant sources of formaldehyde are in wood panels that use glues containing UF resins. These are commonly found in interior grade plywood. Other pressed wood products, such as plywood and oriented strand board, are produced for exterior construction and contain phenol-formaldehyde (PF) resin. Although off-gassing patterns vary based on use of UF or PF, each chemical off-gasses for a significant time. A means of controlling this release of formaldehyde is to seal each sheet with an appropriate low VOC paint or sealer. All surfaces of particleboard should be sealed with either a water-based sealant or a low or no VOC paint or sealant. Another alternative is to select alternate materials that have little or no such chemical.

Low VOC Materials

Low VOC paints, stains, and finishes are one more way to improve your indoor air quality. Traditional paints and stains release low-level toxic emissions sometimes for years after initial application. The source of these toxins is a series of VOCs that up until recently were necessary components for paint. This has become one of the major causes of poor indoor air quality. There are alternatives to paint with VOC off-gassing. Many paint manufacturers offer either low or zero VOC paint made from a water base by using natural raw ingredients such as plant oils, resins and dyes, essential oils, and natural minerals such as clay, chalk and talcum, milk casein, natural latex, beeswax, and earth dyes. Using low VOC or no VOC paint and stains is a cost-effective way to dramatically reduce indoor pollutants. Another option is to provide adequate ventilation for this off-gassing. For instance, the underside of roof decking interfaces with an attic. Because the attic is normally vented, this allows the chemical to dissipate into the atmosphere. However, the decking for a floor or cladding for an exterior wall would normally have no such option, so sealant is the best choice.

Exterior or Exposed Wood

Wood installed in exterior locations or subject to moisture accumulation must either be treated to resist decay or be of a species that would naturally resist decay. The treatments for wood that have enjoyed success for several decades with their roots in ancient history are with the Romans treating ship hulls with tar and the Greeks drenching wood used for bridges in olive oil. Modern treatment using advanced chemicals has been around for about 75 years. In the last few years, a change in the type of treatment used for wood in areas subject to decay reduced the use of chromated copper arsenate (CCA), with the manufacturers voluntarily changing to the use of two waterborne compounds: alkaline copper quaternary (ACQ types A, B, C, and D) and copper azole (CBA-A, CA-B). The main ingredient in ACQ is copper, which has a good track record for preserving wood when installed in outdoor conditions. With the voluntary switch, it is very difficult to get CCA-treated lumber, but it is still important to verify that you have ACQ or a material that is resistant to decay. Decay-resistant material includes heartwood of redwood or western red cedar. Before you buy, check with your local jurisdiction on the material that is acceptable under your local conditions. This is a natural alternative to chemical treatment of wood.

Low VOC Carpet

If carpet must be used, installing low VOC carpet instead of standard carpet is an obvious choice. Much like low VOC paint and stain, low VOC carpet will improve your indoor air quality. Environmentally responsible carpet options each have their own merits and drawbacks. Selection depends on your needs, the room, and its use.

Carpet recovery

The Carpet America Recovery Effort, or CARE, develops market-based solutions for the recycling and reuse of postconsumer carpet products. Among the services it provides is assistance to find carpet reclamation partners in your area. The CARE is a joint industry-government effort intended to increase recycling and reuse of used carpet in an effort to reduce the amount of waste carpet that ends up in landfills. Its membership includes carpet manufacturers.[17]

Carpet durability

If the carpet you intend to purchase is for a high traffic area, durability should be the primary concern. Remember, the durability of any product directly relates to its shade of green. Select colors and patterns that do not exhibit wear. Dark, extremely light, or dramatic colors such as red show soil more easily. Multicolored tweed shows less soil than a solid color. Earth tones also hide soil better.

Another aspect to measure carpet durability is the method of manufacture, its appearance, and performance. Lower pile height and higher pile yarn density are better for high traffic areas and give the best performance. A commercial grade carpet is typically more durable because it is designed to withstand high traffic patterns. Although it may not be the cheapest product, it lasts longer and provides greater serviceability.

Indoor air quality and carpet adhesive

Indoor air quality is adversely affected by adhesives for carpets that emit VOCs. If there is an alternative to glues such as tack strip, use it if possible. Most carpet glues and pads are more toxic than the carpet itself. Avoiding the use of glue with the mechanical means of stretching carpet also does not destroy the floor surface as can be the case with the use of glues.

Health Effects

Lead dust, mercury, asbestos, and radon are all serious threats to health. These elements and chemicals are present in homes in a variety of construction materials. Each affects indoor air quality as well as the health and safety of the occupants of buildings subject to certain conditions and concentrations. Lead is more prevalent in older homes and might be present during remodeling or room additions that are intended to be green. The presence of radon is possible on new homes because it is a naturally occurring gas released from within the Earth.

Lead

Airborne lead and lead dust are particular dangers to infants and small children. Lead can affect physical and mental development and cause acute illness in both children and adults. According to the EPA, "Most health professionals know the threat of lead toxicity, particularly its long term impact on children in the form of cognitive and developmental deficits which are often cumulative and subtle."[18] Lead is most commonly found in paint on the walls of older buildings. Older homes or housing units still contain some lead-based paint.

[17] *Carpet America Recovery Effort, http://www.carpetrecovery.org/*
[18] *U.S. Environmental Protection Agency, http://www.epa.gov/*

The most common means of ingestion of lead by children is inhaling lead dust, not by licking or chewing the painted surface. According to the EPA, "Lead levels in paints for interior use have been increasingly restricted since the 1950s, and many paints are now virtually lead-free. But older housing and furniture may still be coated with leaded paint, sometimes surfacing only after layers of later, non-lead paint have flaked away or have been stripped away in the course of restoration or renovation. In these circumstances, lead dust and fumes can permeate the air breathed by both adults and children."[19] Fortunately, paint and most modern plumbing materials are free of lead. However, if your project involves remodeling an older home, have it evaluated before you make an offer. As with asbestos, the cost of removing this material could be an unwelcome surprise.

Mercury

Mercury vapor off-gassing from latex paint is a relatively new concern determined to be a danger as recently as 1990. Some latex paints have an added chemical called phenylmercuric acetate (PMA). PMA is a preservative that was used to prolong the paint's shelf life. Action taken by the EPA in August 1990 was to eliminate the presence of mercury compounds from indoor latex paints where they were manufactured. Paints containing mercury, including existing stock originally designed for indoor use, were identified with the words "*For Exterior Use Only.*"

Asbestos

Asbestos is found in older homes and buildings. It was once widely used in building materials such as roofing shingles, fireproofing, insulation, heating systems, and floor and ceiling tiles in older buildings. When asbestos-containing material is damaged or disintegrates, microscopic fibers are dispersed into the air. This condition is called *friable.* Although most asbestos-associated cancers are related to the intensity and duration of exposure, the symptoms of the disease do not usually appear until about 20 to 30 years after the first exposure to asbestos. Attempting to remove asbestos-containing materials is not a good idea because the asbestos fibers can be released into the air during the removal process. The EPA requires removal only in order to prevent significant exposure. A management program for intact asbestos-containing materials is often recommended instead.

Fortunately, asbestos is not in most building materials available for home construction. However, if your project involves restoring an older home, have it evaluated before you make an offer. The cost to remove asbestos properly will be an unwelcome surprise.

Radon

Radon is a naturally occurring gas that comes from the Earth and passes into the atmosphere. The EPA estimates that as many as six million homes throughout the country have elevated levels of radon.[20] The EPA and the Office of the Surgeon General have recommended that homes in radon-prone areas, below the third floor, be tested for radon. When a house is built over a source of the emission, the gas enters the building. If the building is relatively airtight, the radon gas will remain and is ingested.

[19] *U.S. Environmental Protection Agency, http://www.epa.gov/*
[20] *Ibid*

According to EPA analysis, Radon is the second leading cause of lung cancer, following smoking. Radon is odorless, colorless, and tasteless. It occurs naturally as a radioactive gas. This comes from the decay of radium, which breaks down into radon and which then becomes airborne and can be ingested into the lungs. The radon decay process produces high-energy alpha particles that increase the risk of lung cancer. The danger of radon has been recognized since the late 1970s. Lung cancer is a disease that is associated with radon exposure. Tobacco smoke in combination with radon exposure has a compound effect. Smokers and former smokers are believed to be at especially high risk from radon. According to Rebecca Morely, Executive Director of the National Center for Healthy Housing, radon leads to approximately 21,000 lung cancer deaths each year.[21]

Testing for radon

Generally, radon does not present itself until the ground has been disturbed or altered. However, when you take that step and begin construction, a short-term test can be used to evaluate the presence of radon, and if needed, plan for getting rid of this naturally occurring, invisible gas. The concentration of radon in the air is measured in "picocuries per liter of air," or "pCi/L." You can test for radon with low-cost radon test kits available in some hardware stores, or you can hire a professional to test for you.

Radon mitigation

Imagine a beach ball underneath a flat bottom boat trying to break through the hull. The beach ball in this analogy is radon and the flat bottom is a concrete slab, trying to hold it back. Radon will enter your home from the earth through any of a myriad of cracks in the slab. There are several methods that can be used to lower radon levels in homes. Radon mitigation is relatively easy during the construction process. Some techniques prevent radon from entering the home, whereas others reduce radon levels after it has entered. If you create an easy path, the radon will follow that route instead. The type of foundation system will guide toward the best method for radon mitigation. For basement or slab-on-grade houses, an effective measure involves actively extracting the gas from the home. Suction pipes are inserted through the slab into the crushed rock beneath. The number of pipes depends on the concentration of radon and how easily gas can move through the rock. Fans can be used to siphon the gas from the pipes (active method).

A passive vent system relies on air currents to extract the gas. According to the EPA, "This system, known as a soil suction radon reduction system, does not require major changes to your home."

To improve on this method, a plastic membrane placed beneath the slab (before concrete pour) will serve to form an additional barrier.

Mitigation in homes with crawl spaces can be achieved by ventilating the crawl space much like conventional foundation venting. This dilutes the radon beneath the house. Adding vents or using powered fans increases the dilution. In addition, covering the soil with heavy plastic sheeting retards the flow of radon into the habitable spaces. The joints of plastic sheets should

[21]*National Center for Healthy Housing, http://www.nchh.org/*

be sealed at the seams and any penetrations, such as plumbing or foundation support, in some manner because radon gas will otherwise seep through. To extract any residual gas beneath the plastic membrane, use a fan to draw radon out and away from the house.

BEFORE YOU DECIDE . . . REFLECTIONS AND CONSIDERATIONS

- ✔ Indoor air quality is based on biological and chemical factors as well as lifestyle, equipment maintenance, and moisture conditions.
- ✔ Mechanical equipment that moves air must be maintained properly through regular cleaning.
- ✔ Moisture control is essential to indoor air quality.
  - Design the building to avoid damage due to rainfall and infiltration with several measures:
    - Elevating the finished floor surface
    - Having capillary breaks between porous materials and the soil
    - Elevating siding and structural members above moist conditions
    - Adding gutters and downspouts
  - Monitor any equipment that includes transfer or containment of water to avoid unnecessarily damp conditions.
  - Use mold-resistant materials.
- ✔ Ventilation is essential to indoor air quality.
  - Exhaust fans for the bath, dryer, and range remove harmful fumes, vapors, and odors from habitable space.
  - Properly conditioned airflow delivery is essential.
    - Seal newly installed ducts during construction to prevent accumulation of debris or construction dust.
    - Use ceiling fans to increase air circulation and to prevent stagnant air within your home.
  - The installation of a central vacuum provides convenient cleaning while keeping all the pollutants outside for storage awaiting disposal.
  - A blower door fan test will tell you what kind of job you have done in sealing air leaks.
  - Use caution regarding materials with VOCs that cause adverse health effects.
    - Chemicals including mercury, lead, and radon all adversely affect indoor air quality.
    - Avoid use of products with harmful VOCs that off-gas.
      - Use low or no-VOC materials.
    - Seal products containing VOC with paint or similar materials.

FOR MORE INFORMATION

Carpet America Recovery Effort
http://www.carpetrecovery.org/

EPA 2003 Annual Report
http://www.epa.gov/ocfo/par/2003par/2003ar.pdf/

EPA Energy Star
http://www.energystar.gov/

EPA IAQ Building Education and Assessment Model (I-BEAM)
http://www.epa.gov/

EPA Indoor Air Pollution
http://www.epa.gov/iaq/pdfs/indoor_air_pollution.pdf/

EPA Radon
http://www.epa.gov/radon/healthrisks.html/
http://www.epa.gov/radon/pubs/citguide.html/

FEMA
http://www.fema.gov/

Green Guidelines to Protect Residents from Health Hazards in Their Homes
http://www.nchh.org/

National Center for Healthy Housing
http://www.centerforhealthyhousing.org/

Chapter 7

ENERGY CONSERVATION AND EFFICIENCY

ENERGY MANAGEMENT AND GREEN HOMES

Energy conservation and efficiency are essential in a green home. Efficient use and savings of all energy define sustainable design. The less we use now, the more there will be available in the future. Wasteful energy practices can be changed to bring about net energy savings. This represents the low-hanging fruit in energy management. President Lyndon Baines Johnson (LBJ) earned the moniker *Light Bulb Johnson* when he turned off lights in unattended rooms in the White House. There are examples of conservation from earlier times. With the energy crises in the mid-1970s, conservation received more attention.

In 1974, the Energy Research and Development Administration was created and led to the Department of Energy Organization Act, which led to the creation of the Department of Energy in 1977. That department developed a comprehensive energy plan for the nation. They researched and developed energy technology and conservation and many other programs. This led to a Model Energy Code in the 1980s and an Energy Conservation Code in the 1990s. During this time, advocates for sustainable design had made inroads into social consciousness with the United States Green Building Council introducing its Leadership in Energy and Environmental Design standard for commercial buildings. At the same time, the three model code organizations—Building Officials and Code Administrators (BOCA), International Conference of Building Officials (ICBO), and Southern Building Code Congress International (SBCCI)—reached a decision to merge and became the International Code Council (ICC) Two of its first publications were the International Energy Conservation Code (IECC) and the International Code Council Performance Code for Buildings and Facilities.

ENERGY CONSERVATION BASICS

Energy conservation includes the appropriate use of natural resources by conserving the energy used while still maintaining a comfortable environment. An example of this is setting the thermostat to a temperature that is comfortable yet allows the building systems to maximize their use of energy. Designing and building a house with energy conscious measures can be challenging. There are many methods available to maintain a comfort level in your home while using minimal amounts of energy. The most common techniques include adding more insulation, creating a tight thermal envelope, and monitoring ventilation.

Thermal Envelope

The thermal envelope includes everything that seals the interior environment from the outdoors. At the extreme, a thermal envelope house design uses a principle of complete isolation of the insulated wall, floor, and roof systems within a habitat. This type of house design originated as a result of the energy crises of the 1970s. It is an architectural design known as the double envelope with twin walls separated by an air space that retards heat loss through conduction or convection. Although the design performed well, it proved that there are more considerations in a home besides heat loss. Convection of heat was not uniform throughout the house and heat seemed to be stratified. The air barrier created a chase in the walls that is prohibited in conventional framing. The chase creates a chimney effect that would lead to the rapid spread of fire and loss of the building and perhaps human life. Controlling the spread in the chase is done with fireblocking that is installed to retard the passage of heat, smoke, and fire between stories and along wall spaces over 10 feet in length. The Code lists the type of materials required.

> *602.8.1 Materials.*
>
> *Except as provided in Section R602.8, Item 4, fireblocking shall consist of 2-inch (51 mm) nominal lumber, or two thicknesses of 1-inch (25.4 mm) nominal lumber with broken lap joints, or one thickness of 23/32-inch (19.8 mm) wood structural panels with joints backed by 23/32-inch (19.8 mm) wood structural panels or one thickness of ¾-inch (19.1 mm) particleboard with joints backed by ¾-inch (19.1 mm) particleboard, ½-inch (12.7 mm) gypsum board, or ¼-inch (6.4 mm) cement-based millboard. Batts or blankets of mineral wool or glass fiber or other approved materials installed in such a manner as to be securely retained in place shall be permitted as an acceptable fire block. Batts or blankets of mineral or glass fiber or other approved nonrigid materials shall be permitted for compliance with the 10 foot horizontal fireblocking in walls constructed using parallel rows of studs or staggered studs. Loose-fill insulation material shall not be used as a fire block unless specifically tested in the form and manner intended for use to demonstrate its ability to remain in place and to retard the spread of fire and hot gases.*[1]

Although a solution is to install sufficient batts of fiberglass insulation in these interstitial areas to act as fireblocking, it is a resource-intensive and costly approach to creating an energy-efficient thermal envelope.

Life-Cycle Assessment

There is an environmental impact made by energy conservation measures. Although we achieve a reduction in overall energy used, it comes at a cost, albeit relatively small. In order to achieve energy conservation, we have developed materials and processes that demand energy themselves. Some of those innovations consume more energy overall than what is saved in the lifetime of the

[1] *2006 International Residential Code, International Code Council, Inc., Section 602.8.1*

home. Consider any energy conservation measure and you will see an associated energy demand in the life cycle that includes production, delivery, and installation of that product that otherwise would not be present. The goal of a life-cycle assessment is to compare the environmental effect of products and services to be able to choose the one with the least environmental impact. It includes everything from the harvesting or mining of raw material and its production, manufacture, distribution, stocking, delivery, shelf life, use, and disposal. You can use this evaluation between products to evaluate the comparative environmental impact.

For example, a well-regarded method of wall construction, called insulated concrete forms (ICF), uses rigid synthetic insulation as formwork for poured concrete. This forms a reinforced concrete wall with insulation on both sides. This type of wall system saves energy for heating and cooling because of a higher R value and a tighter wall system. This product had to be designed, developed, and manufactured then delivered to a distributor and finally to the project site. This process called for the use of energy that would not have otherwise been expended. Additionally, the solid waste production and air pollution caused by the manufacture and delivery of all materials, including concrete, are all part of the life-cycle assessment that determines the true nature of the green home material or component.

EXCEEDING THE CODE

Meeting the Energy Code should be regarded as the starting line for a green home. The energy provisions of the International Residential Code (IRC) establish a baseline threshold for minimum requirements of energy efficiency. The Energy Code, while providing a higher level of energy conservation than previously achieved, still represents the low end of the scale for a truly green home. It should be expected to exceed the Energy Code for your green home. The level of exception is your choice. For one- and two-family houses, the pertinent Code provisions are in Chapter 11 of the IRC.

Prescriptive Compliance

The IRC, Section N1101, allows a choice between meeting the prescriptive standards of Chapter 11 or the performance standards of the International Energy Conservation Code. To begin, find the climate zone for your area from Figure N1101.2 or Table N1101.2. Consider a conventional wood frame house with simple pitch roof trusses. The floor is a monolithic concrete footing/slab. The building is located in Luna County, New Mexico, which falls under Zone 3. Use Table N1102.1 to determine the prescriptive minimum type of insulation needed. A maximum U value of 0.65 for all fenestration, including skylights, is needed for glazing. But for glazed fenestration you need a maximum of 0.40 solar heat gain coefficient (SHGC). The attic or roof must be insulated at R-30, and the walls at R-13. Ducts supplying conditioned air must be insulated with R-8. If you use paper-faced insulation, the required vapor barrier is provided as long as it is installed properly. Air leakage is regulated for about eight specific conditions. The slab edge does not need to be insulated.

Energy Code Checklist

When you complete your energy efficiency installation, a building inspector will verify these basic installations. Following is a list of questions, followed by the standard for inspection, to use as a checklist. An inspector will normally look for these items after insulation has been installed.

1. *Are proper materials installed?* Only approved materials may be used as insulation materials. These are manufactured type with R- value printed on each piece.
2. *Are all windows and doors installed according to the standards required on the approved plan?* Windows and doors must meet the energy standards for your climate zone. They will be identified with a label indicating a U value. In addition, the air leakage rates on the label must match those required. The IRC includes energy standards for windows and doors.

   Fenestration air leakage. Windows, skylights and sliding glass doors shall have an air infiltration rate of no more than 0.3 cubic foot per minute per square foot [1.5(L/s)/m²], and swinging doors no more than 0.5 cubic foot per minute per square foot [2.5(L/s)/m²], when tested according to NFRC 400 or AAMA/WDMA/CSA 101/I.S.2/ A440 by an accredited, independent laboratory, and listed and labeled by the manufacturer. Check this against the approved plans.[2]
3. *Is there an insulation certificate available (or required) to certify compliance with installation?* The installer must leave you a certificate indicating that the insulation has been completed. This certificate will indicate the type and value of the insulation.
4. *Are all exterior walls insulated?* Verify that the walls are insulated properly according to your climate zone.
5. *Is the ceiling or attic insulated?* Verify that the walls are insulated properly according to your climate zone.
6. *Is the floor or crawl space insulated?* Verify that the walls are insulated properly according to your climate zone.
7. *Is vapor barrier installed?* A vapor barrier is required to be installed on the warm (or heated) side of the insulation. Vapor retarders limit the amount of moisture vapor that passes through a material or wall assembly. The moisture transmission rate of a material is referred to as *permeability* and is stated in terms of *perm inches*. Divide the permeability of a material by its thickness for the material's permeance (stated in perms). Permeance is the number that should be used to compare various products with regard to moisture transmission resistance.
8. *Is cross-ventilation provided in the attic and crawl space?* Cross-ventilation is required in attics and crawl spaces at the rate of 1 square foot for every 150 square foot of attic or floor space it serves.
9. *In the attic, are baffles provided at soffit ends to facilitate required ventilation?* To facilitate ventilation on eave or cornice vents, a 1-inch clearance is required. Baffles are one approved way to help provide this.
10. *Is the flexible duct for supply and return air insulated to R-5?* Inside ductwork located in unheated spaces must be insulated to R-5. Outside ductwork must be insulated to R-8.

Performance Path

The performance path through the International Energy Conservation Code is another method of compliance with a minimum standard, but it requires a few more steps. The value in using this approach is that sometimes a design would not be able to meet the strict requirements of the prescriptive path.

[2] *2006 International Residential Code, International Code Council, Inc., N1102.4.2*

N1101.2 Compliance.
Compliance shall be demonstrated by either meeting the requirements of the International Energy Conservation Code *or meeting the requirements of this chapter.* This allows for the option of using the IECC to establish compliance. Since you're obviously going to exceed the code, the rationale for using the performance path may be because the type of construction you plan to use does not appear in the IRC as a commonly used material. Therefore you need to establish thermal resistant factors equivalent to those R values located in either the IRC Table N1102.1.2 or the IECC.[3]

R Value and μ Value

The letter *R* is the comparative value of an insulation's ability to resist the flow of heat. The higher the R value, the more that insulation can resist the flow of heat. The Greek letter μ is the inverse of R or 1/R. As the material's μ value decreases, the more resistance that material has to the flow of heat. The smaller the μ value, the higher the R value. Generally, insulation is measured in R value and windows or doors are measured in μ value. For example, a window with a μ value of 0.40 is more thermally resistant to heat flow than a window with a μ value of 0.65.

Additionally, the μ value is important for calculating the precise heat loss of a gross wall section with components that have differing R or μ values. For instance, consider the house in Luna County, New Mexico, (Climate Zone 3) that has an exterior south-facing wall. The length is 65 feet 6 inches and height is 8 feet (524 square foot gross wall). The wall has five windows for solar heat gain purposes. Each window is 20 square feet or 100 square feet total glazing and rated with a $\mu = 0.60$.

There is one wood frame door that is 3 feet by 6 feet 8 inches and has a rated μ value of 0.60. The walls are wood frame and insulated with R-19 insulation.

PROBLEM: Determine the (μ) × (Area) for each component in the wall. This way, you can validate your compliance with Table 402.1.3 in the IECC.

Step 1. Gross wall area – windows – door = Net wall area
Net Wall Area = 524 ft^2 − 100 ft^2 − 20 ft^2 = 404 ft^2 @ R-19 or $\mu = 0.053$

Step 2. Calculate μ A for net wall area:
With R-19, $\mu = 0.053$: therefore, $\mu A = (404 \text{ ft}^2) \times (0.053) = 21.412$

Step 3. Calculate μ A for net window area:
With $\mu = 0.60$: $\mu A = (0.60) \times (100) = 60$

Step 4. Calculate μ A for net door area:
With $\mu = 0.60$: $\mu A = (0.60) \times (20) = 12$

Step 5. Calculate overall μ A for overall wall section: $21.412 + 60 + 12 = 93.4$

Step 6. Compare with the standard set out in Table 402.1.3:
$(0.65) \times (100) + (0.65) \times (20) + (524) \times (0.082) = 65 + 13 + 43 = 121$

[3] *2006 International Residential Code, International Code Council, Inc., Section N1101*

According to Section 402.1.4 of the IECC:

> *If the total building thermal envelope UA (sum of U-factor times assembly area) is less than or equal to the total UA resulting from using the U-factors in Table 402.1.3 (multiplied by the same assembly area as in the proposed building), the building shall be considered in compliance with Table 402.1.1.*[4]

Since 93.4 is less than the design minimum set out in the Table 402.1.3,you have surpassed the Code; in this case by about 20%. This is an accurate method to determine your comparative improvement over the base standard.

R Value and Wall Assembly

There are conditions where placement of the insulation in relation to the conditioned space changes the paradigm normally accepted regarding heat flow. Two wall assemblies with identical R values can have different thermal performance features. For instance, a concrete masonry unit (CMU) block wall with insulation inside the thermal mass of the wall does not perform depending on the direction of heat flow, in the same manner as a similar CMU wall with insulation on the outside. Most energy modeling programs do not account for these variables.

Exception to the General Rule

Another exception to the principle regarding the significance of insulation values as it relates to retarding heat flow is adobe construction. Adobe is earthen material that behaves differently than an insulated cavity or masonry wall or even a (noninsulated) solid masonry wall. It is like a combination of the two. It relies on the thermal mass and the time lag for heat flow. For instance, an adobe home may have no insulation, yet maintain a comfortable temperature during a time when heating would otherwise be necessary. The reason is that by the time the heat transfers through the thermal mass of the wall is reversed because of nightfall and then reaches the outside, morning brings sunshine and a renewal of the cycle. However, this only works in climates with large enough temperature swings from day to night. In extended very cold or hot periods, thermal mass will become cooler or hotter than desired.

■ CONTROLLING HEAT LOSS THROUGH AIR SEALING MEASURES

Air leakage sealing measures are crucial to energy conservation and efficiency. These sealing techniques prevent uncontrolled leakage of conditioned air that you have purchased through energy usage. These leaks occur in penetrations, gaps, or joints in the fabric of the building and represent heat loss. The goal is to tighten up the thermal envelope everywhere within the house.

EPA Energy Star Thermal Bypass Checklist

Energy Star qualified homes must pass a thorough checklist of energy-saving practices. This checklist of items includes the following major categories:

- Overall air barrier and thermal barrier alignment
- Walls adjoining exterior walls or unconditioned spaces

[4] *2006 International Energy Conservation Code, International Code Council, Inc.*

- Floors between conditioned and exterior spaces
- Shafts
- Attic and ceiling interface
- Common walls between dwelling units[5]

This step ensures that the Energy Star qualified home will have improved energy efficiency. The thermal bypass checklist is relatively new and serves as a voluntary standard for those builders who wish to participate in the program. The checklist provides similar ways to seal thermal leaks and energy drains in a typical house.

Testing the Tightness of the System

Building envelope air leakage tests the air changes that occur per hour using a standard blower door testing protocol set forth by the American Society for Testing and Materials (ASTM). A blower door test is an appropriate method to detect leakage in this type of system. The test uses equipment like that shown in Figure 7-1 that pressurizes the interior of the home. The test must indicate less than seven air changes per hour when tested with 50 pascals of pressure to meet this blower door test requirement in the 2009 IRC.

Foundation

The foundation wall of the building should be solid with no penetrations except for those required for foundation ventilation and flood venting. The foundation, if masonry or concrete, should have steel anchors to connect to the sole or sill plate. The connection between the foundation and the anchor should be filled with caulking that will seal the joints that occur between the dissimilar materials.

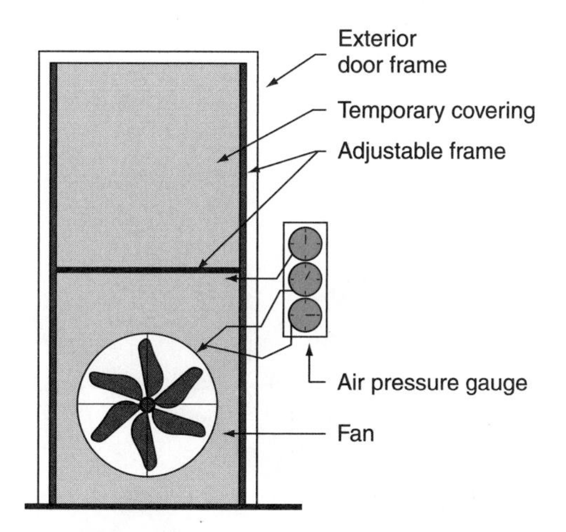

Blower door diagnostics determine building envelope leakage.

Courtesy of Southface/Earthcraft House

Figure 7-1 ■ Building envelope tightness is evaluated with a blower door test. This verifies the air tightness of the building and checks for any air leakage.

[5]*Energy Star, http://www.energystar.gov/*

Floor Framing

Floor framing represents a common place for penetrations of the thermal envelope. The joint between the floor framing and the foundation plate should be sealed with foam board, caulk, or tape. For a floor joist that connects to a foundation plate, the normal connection is a toenail from each side (because the joist is resting on its end). Most commonly, the joist ends 1½ inches from the outside edge of the foundation wall so that the band joist rests fully on the foundation plate. Both of these elements should have a thermal seal that prevents air leakage.

Penetrations

A penetration represents a source for a thermal leak. Conditioned air will find such a leak and escape from a warm home. Alternately, in the cooling season, hot air will seek a penetration leak to enter the conditioned space. These penetrations must be sealed or eliminated.

- *Floor Penetrations*: Floor penetrations between unconditioned and conditioned space include all penetrations, such as bored holes and notches in the floor framing assembly. This would be caused by pipe, wire, ducts, and similar penetrations. The annular space around these penetrations may be sealed with a wide variety of sealant, including caulk, insulation, and foam tape in order to establish an airtight seal.

  Cantilevered floors such as the one shown in Figure 7-2 should be sealed above supporting walls or floors with conditioned space cantilevered over outside areas. The space between should be insulated and then sealed with sheet material or blocking and sealant above the top plate of the supporting wall.

- *Ceiling Penetrations*: The ceiling in a conditioned space is basically a shell, consisting of the bottom of trusses or joist framing that is covered with drywall. There should be a sufficient amount of insulation. However, this insulation is meaningless if there are unsealed leaks in the diaphragm. Penetrations of the ceiling membrane should be sealed between conditioned and unconditioned spaces.

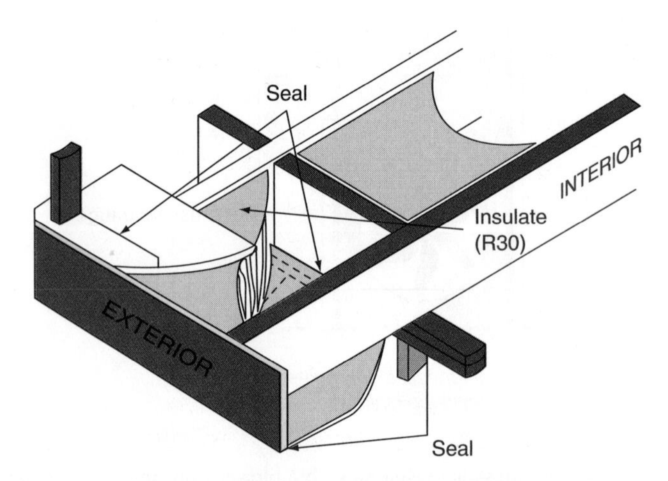

Air seal and insulate cantilevered floors.

Courtesy of Southface/Earthcraft House

Figure 7-2 ■ A cantilevered floor is one that projects out away from a supporting element such as a wall. The area under the cantilevered floor must be insulated because it is exposed to unconditioned, outside conditions.

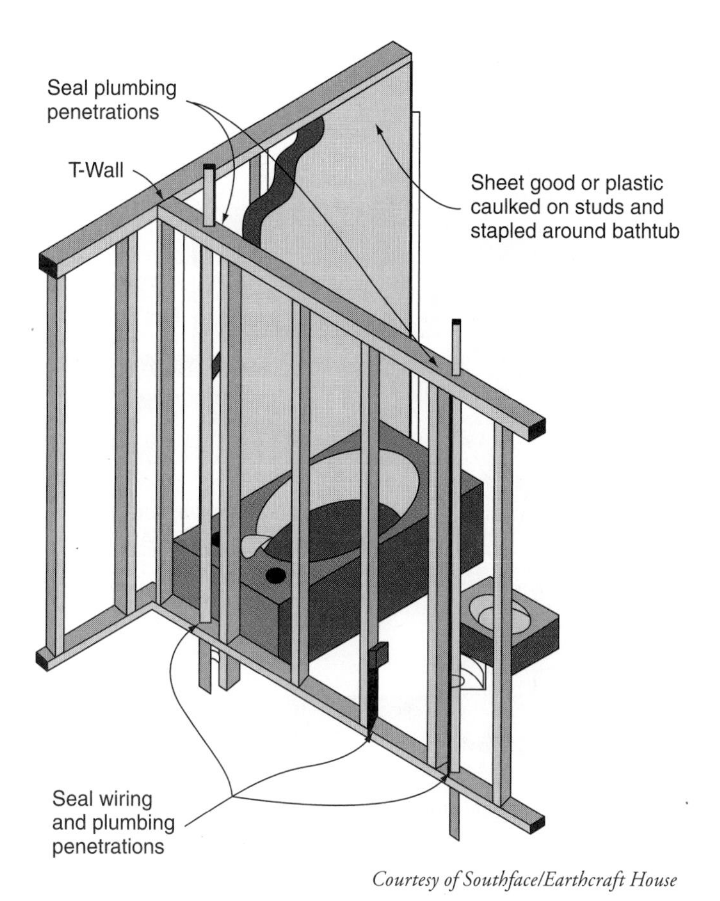

Courtesy of Southface/Earthcraft House

Figure 7-3 ■ Bathrooms are particularly difficult to seal airtight. Plumbing pipe and wiring of hydromassage motors both must transition through walls and floors, leaving gaps in the thermal envelope that must be sealed.

- *Penetrations at Bathrooms*: Bathtub and shower drains like those shown in Figures 7-3 and 7-4 can cause openings in the subfloor between conditioned and unconditioned areas. These openings should be sealed with sheet material and sealant. If the fixture is near unconditioned space, the opening should be insulated and covered with material such as plastic, drywall, or wood sheathing.
- *Framing Penetrations*: The perimeter band joist that connects conditioned floors should be sealed at the top and bottom edge with foam tape or caulk. This is an area that can easily be ignored during framing and cause significant leaks.

Wall framing, like floor framing, is a common place for penetrations in the thermal envelope to occur. There are so many ways to build a wall frame. Many of them are well accepted and meet the Code but they do not provide the means to make a tight house. That requires some additional planning and corrective measures. The perfect time to find and seal potential energy leaks through penetrations is before installing insulation and drywall.

Drywall is penetrated in all walls for a variety of reasons, including plumbing pipe and fixtures, electrical boxes, mechanical boxes, and ductwork. In all of these instances, a tight seal can be attained with caulking, adhesive, or foam tape. It is better to make this seal from the back side of the drywall if done with tape in order to avoid surface irregularity.

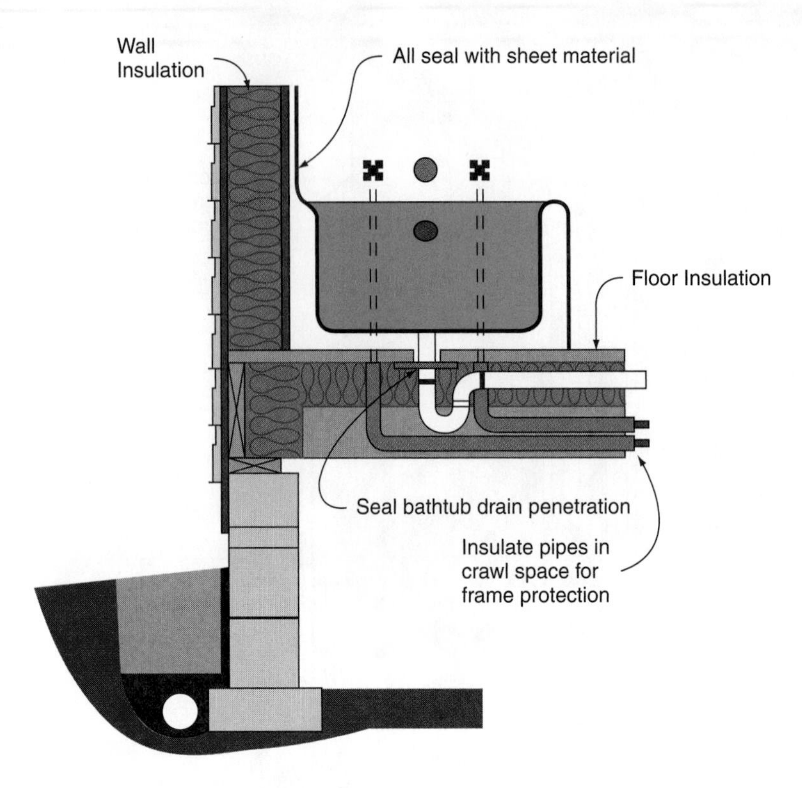

Courtesy of Southface/Earthcraft House

Figure 7-4 ■ Holes created by the bathtub drain must be sealed. There are numerous penetrations in drain and water supply pipes that are in the walls or floor. These are avenues for energy loss.

Bottom and Top Plates

The bottom plate of the perimeter of all exterior wall frames should be sealed to the subfloor with caulk, spray foam, gasket, or construction adhesive. An example of the need for sealing a bottom plate is illustrated in Figure 7-5. Sealing the plate with caulking or gasket forms a lasting barrier for air leakage. Sealing includes spreading beads of latex caulking before erecting the wall frame in place or installing some gasket material on the bottom of the plate.

The joint between the drywall and the subfloor should also be sealed to the bottom plate of exterior walls. This joint, illustrated in Figure 7-6, can easily be sealed with a bead of caulking along the bottom of the drywall before it is installed or to the inner face of wall bottom plates.

The top sheet of drywall on wall frames must be sealed to the top plate of the wall frame. Although many of these are nailed, the connection is important because it tightens the seal from the top plate down through the outside of the wall stud frame. The only layer between the unconditioned air and the interior is the drywall.

Fireplaces and Chimneys

Fireplaces are most commonly factory made and placed within wood frame walls. When they are installed on exterior walls, the exterior wall stud spaces should be covered with sheathing

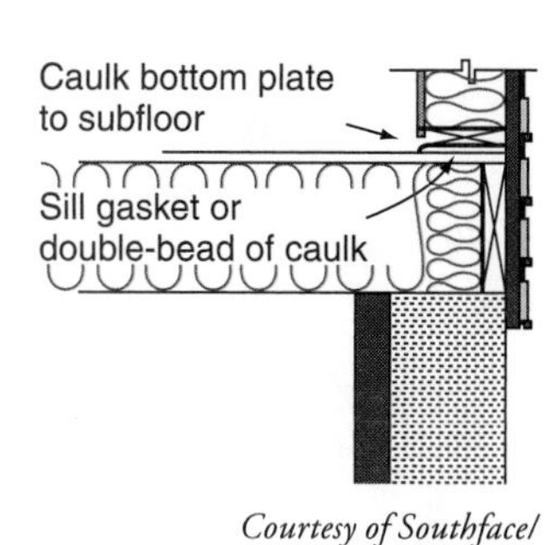

Courtesy of Southface/ Earthcraft House

Figure 7-5 ■ In a wooden wall frame, the lower structural element is a bottom plate. It connects with a floor joist or foundation wall. The joint is prone to air leakage. Gasket installation and caulking are simple ways to plug the leak.

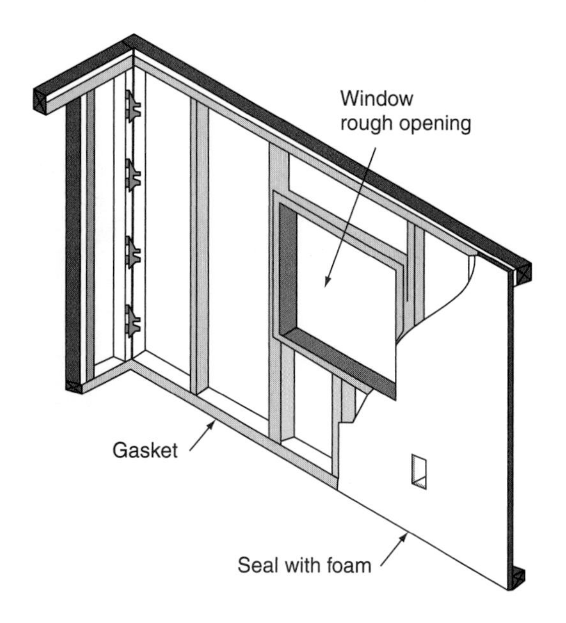

Gaskets, caulk, or foam can be used to air seal drywall at any stage of the installation.

Courtesy of Southface/Earthcraft House

Figure 7-6 ■ Another area to seal is the joint between the drywall and the bottom plate. Normally nails or screws secure the assembly, but these still allow heat transfer. Seal with a gasket or foam before fasteners are installed.

or drywall prior to installation. The annular space around flue penetrations should be sealed with noncombustible sheet material and high-temperature sealant that meet the conditions of the chimney manufacturer.

Exterior Sheathing

Siding is the exterior skin of a building. It is in this critical area where significant leaks occur because of penetrations, joints, seams, and lapping of all building materials. There are numerous ways to create the leak or crack where heat can escape. It is essential to find each condition and fill the air break with sealant that is effective in plugging the leak.

Exterior wall sheathing should be sealed at the plates and seams. Install caulk, foam tape, or gaskets on the building wrap or sheathing to create an airtight connection to the framing. All holes and penetrations should be effectively sealed with caulk.

Exterior siding covered in house wrap, sealed or unsealed at the plates, seams, and openings, is an effective way to stop heat loss through walls. Unsealed wrap should be installed per manufacturer's specifications, installed in a continuous manner, and should cover the top plate of exterior walls, rough openings for windows and doors, and band joist areas. To form an effective air barrier, house wrap should have all the laps and margins taped or sealed.

House wrap like that illustrated in Figures 7-7 and 7-8 should be installed per manufacturer's specifications and applied in a continuous manner. It should be sealed with tape or other

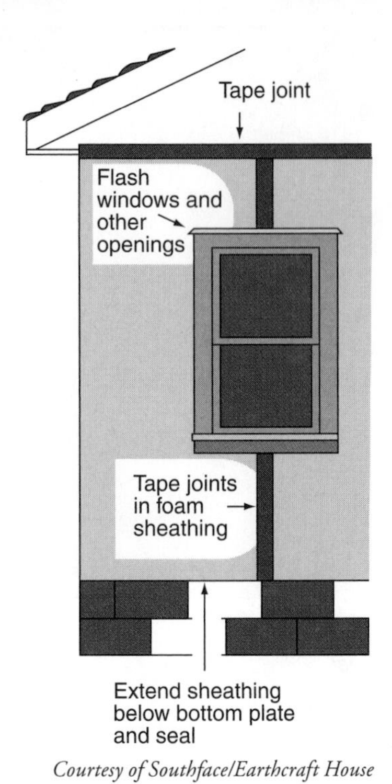

Courtesy of Southface/Earthcraft House

Figure 7-7 ■ House wrap is an effective air barrier for limiting heat transfer from heated areas of the home. Verify instructions before installing and ensure proper lap and installation.

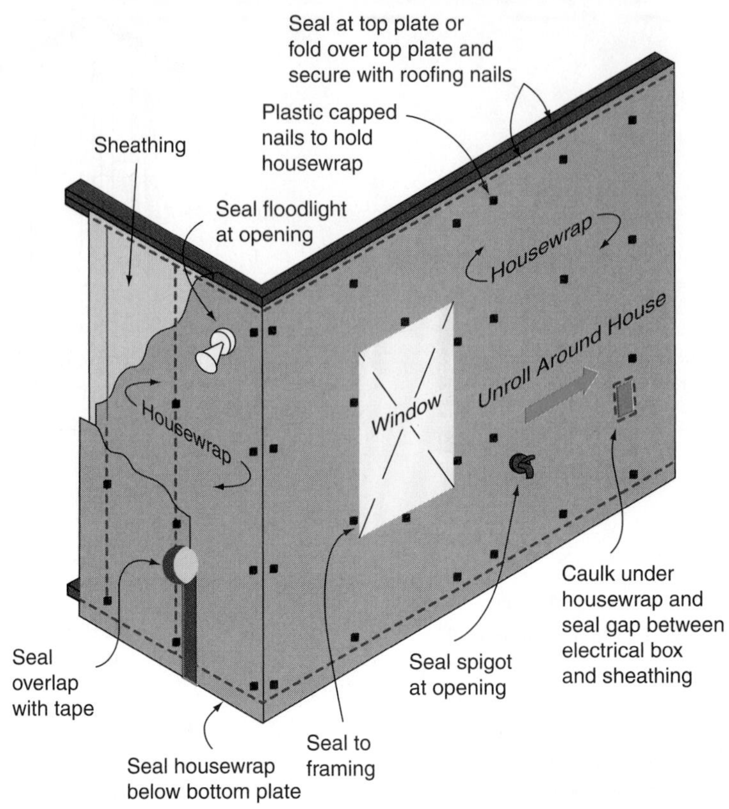

Courtesy of Southface/Earthcraft House

Figure 7-8 ■ House wrap is a manufactured product and must be installed according to certain standards. Seal with overlap and tape as specified by the manufacturer's instructions.

sealant to the framing at the top plates, all overlapped seams, rough openings for windows and doors, all penetrations, band joist areas, and at the foundation wall below the bottom plate.

Roof and Attic

Roof framing is much like floor framing and has similar joints that leak conditioned air. At the point of connection of any framing mechanism with another, a leaky joint can occur. This leaky joint will present itself in a variety of ways. Essentially, every material connection or dissimilar building material is a likely source for energy loss.

Attic Access Panel

Access to an attic is required by the Code.

> *Buildings with combustible ceiling or roof construction shall have an attic access opening to attic areas that exceed 30 square feet (2.8 m²) and have a vertical height of 30 inches (762 mm) or more. The rough-framed opening shall not be less than 22 inches by 30 inches (559 mm by 762 mm) and shall be located in a hallway or other readily accessible location. A 30-inch (762 mm) minimum unobstructed headroom in the attic space shall be provided at some point above the access opening. See Section M1305.1.3 for access requirements where mechanical equipment is located in attics.*[6]

[6]*2006 International Residential Code, International Code Council, Inc., Section R807*

Attic openings must have tight-fitting connections to prevent wholesale heat loss. Attic access is similar to an exterior door. The door itself should be an insulated type. With a loose-fitting connection into the frame and no weather stripping, energy loss would be great.

Attic Knee Walls

A knee wall, like the ones shown in Figure 7-9 are a way to increase habitable space without increasing the size of the home. They are less than full-height vertical stud framed walls within the attic space. This forms a habitable area within the attic by defining a short wall section along the lower slope of the roof line. Attic knee wall doors must be weather stripped and tight fitting with sealed exterior sheathing, similar to that of the attic door. The other side of the wall is unconditioned space and draws heated air that would otherwise stay inside. In Figure 7-9, a knee wall is being insulated, forming a thermal envelope around the habitable space.

Chases

Chases are framed shafts in houses whose purpose is to protect ductwork or air transfer. Some chases are designed and built for a purpose, such as a chimney. Others are inadvertent, leftover space that would not work without being closed up. But, in some cases, chases have a direct link to an unconditioned space, such as the attic or the crawl space. These must be sealed and insulated to prevent heat from passing through.

Courtesy of iStock Photo

Figure 7-9 ■ Habitable attics must be insulated and heated. Knee walls are the short wall segments from the floor to the sloping element of the rafters. These must be insulated to prevent heat loss.

Roof Underlayment

Under your roofing material is a thin film of material that is the last barrier for weather protection. The roofing manufacturer's installation instructions will most likely call for specific sealing and underlayment measures. This is important to observe because the warranty for the roofing is predicated on this installation. Be sure to follow the manufacturer's instructions by following the measures completely and sealing joints in the underlayment.

Insulation

Insulation is the bread and butter for energy conservation and efficiency. The operational characteristic for insulation is to create a dead air space. Energy that moves through convection is carried by moving air. If you slow or stop the air, you stop energy flow. The higher the R value of the insulation, the slower the natural airflow becomes, thereby lowering the transmission of heat.

Concrete and Masonry Insulation

Foundation walls like the one shown in Figure 7-10 and concrete slab floors can lose heat to conductive loss even though airflow is stopped. Insulating and sealing of the leak are necessary. A thermal break is necessary. An example of a thermal break in a slab is rigid insulation between the ground and concrete.

Concrete slabs rest directly on soil. As such, they form a path for heat transfer. They should be insulated to retard this loss of heat. Slabs should be insulated with at least an R-6 (or higher) edge insulation such as foam that could serve as a concrete form. Alternatively, nonmonolithic slabs may use rigid insulation between the stem wall and the slab.

For basements with masonry or concrete walls, install furring strips then rigid insulation in between and cover with drywall from floor to ceiling, observing the sealing measures for the

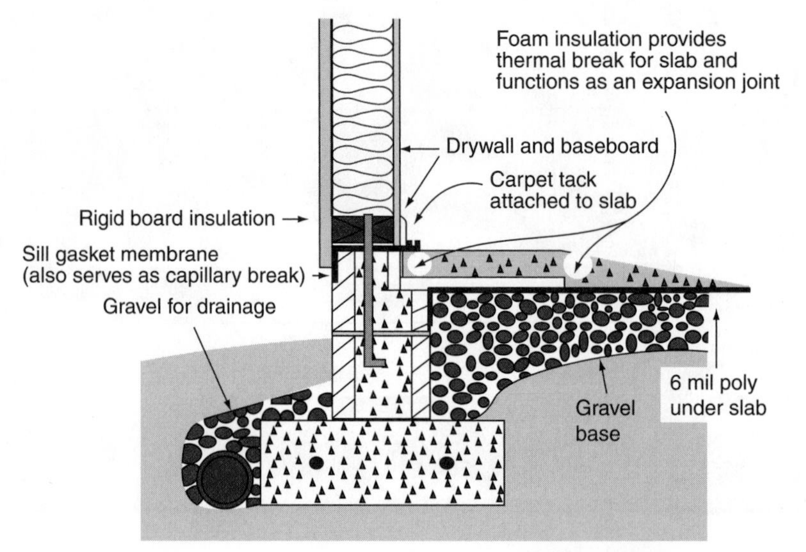

Courtesy of Southface/Earthcraft House

Figure 7-10 ■ The foundation wall must be insulated to retard heat loss. Install furring wall and insulate with solid insulating panels. Other joints, such as floor joist, band joist, bottom plate, and wall frame must be thoroughly sealed to retard air movement.

wall frames above. Areas above the foundation wall should be thoroughly insulated and gaps and seams thoroughly sealed. Caulk the band joist to the subfloor and insulate. Under the wood plate, install a sill gasket or a double bead of caulking.

Ground Floor Insulation

For framed floors over conditioned space, be sure to carry the wall insulation through the space between walls against the band joist. Additionally, caulk the bottom plate of the upper wall to the subfloor. This area will be forever concealed but it is a commonly neglected area that could yield cold drafts and heat loss.

Crawl space insulation

It is important to seal crawl space walls with caulk, foam sealant, or gasket between the foundation stem wall and the sill plate, between the sill plate and the band joist, and between the band joist and the subfloor. These are likely areas for energy loss through air leakage. All penetrations in the crawl space wall such as the one illustrated in Figure 7-11 should be sealed and access doors should be weather-stripped. Although necessary in some climates and geographic locations, openings should be limited to provide for foundation ventilation and flood vents. In addition, foundation walls below grade should have systems that retard moisture accumulation. Notice the perforated pipe near the footing in Figure 7-12. This pipe will collect and remove surface drainage that drains around foundation walls.

Insulating Fireplaces and Chimneys

A manufactured fireplace must be installed according to the manufacturer's instructions. Read them carefully and, if permitted, fill the wall frame with insulation to prevent another common leak. Use at least an R-13 insulation to fill both wall and roof and ceiling assembly as

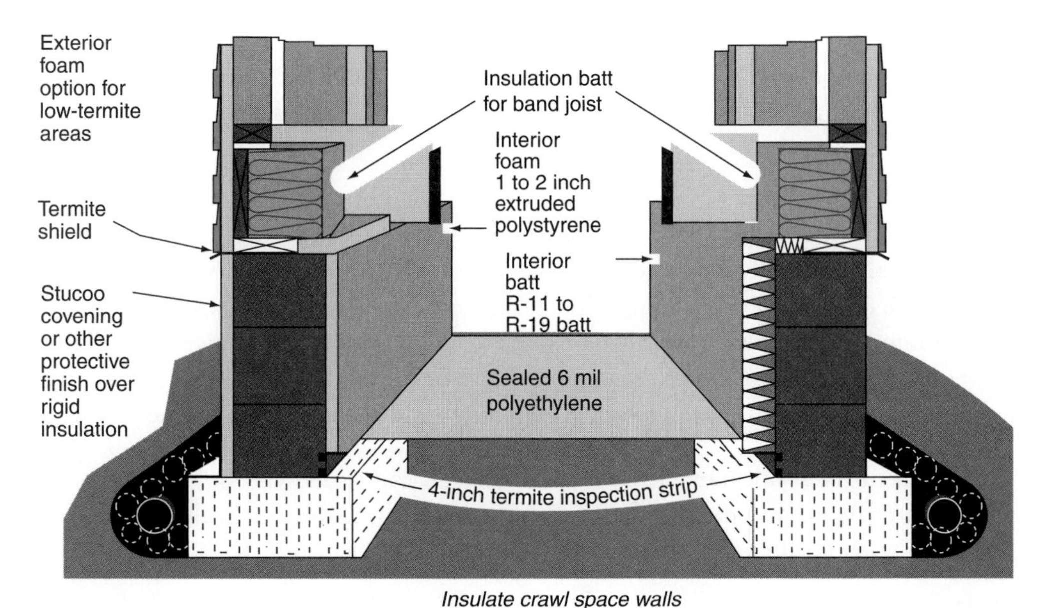

Courtesy of Southface/Earthcraft House

Figure 7-11 ■ A crawl space is the area under a home within the confines of the foundation wall. This area is typically unheated because it interfaces with outdoor conditions. It should be insulated and sealed to retard heat leakage.

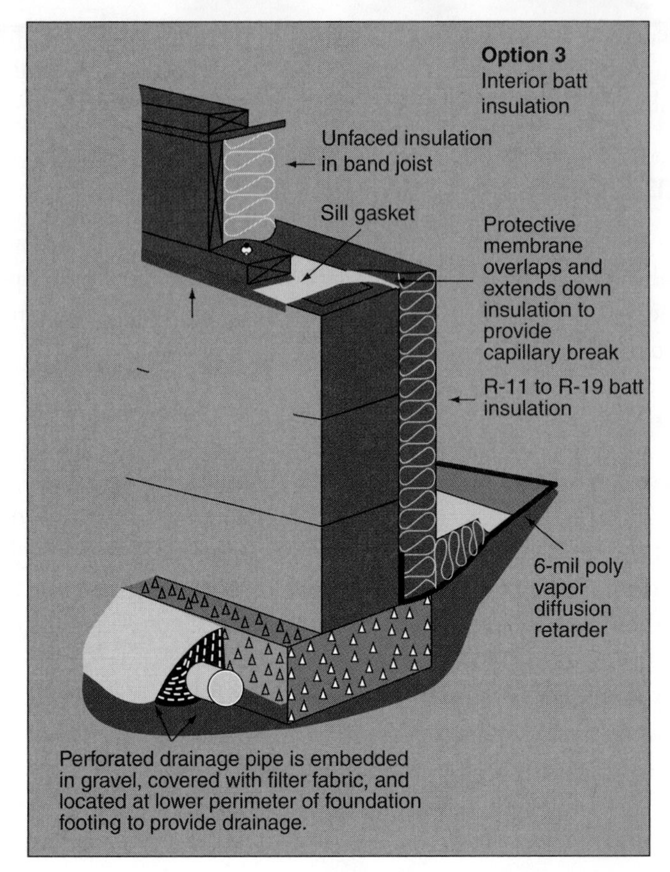

Lynn Underwood

Figure 7-12 ■ A portion of the foundation wall will be beneath the finished grade and can be affected by drainage from precipitation. A drainage pipe embedded in the gravel bed along the footing can remove this water and prevent damage to the foundation. A foundation wall, where it defines an interior crawl space, should be insulated.

illustrated in Figure 7-13. Be careful to hold insulation away from the chimney exhaust. Note that the chimney manufacturer specifies a setback distance for combustible materials.

Wall Frame

Exterior walls, including stud cavities, must be insulated as illustrated in Figure 7-14. The higher the insulation value, the more resistance to heat flow. Fill the wall with the insulation designed for that frame. Do not compress the insulation. Insulation stops the flow of air. Compressing it into the cavity will leave a gap or space that allows air to convectively transfer heat. Most insulation is made for either a standard 4- or 6-inch wall frame. For instance, you can install R-22 batt-type insulation in a 6-inch wall frame that has studs at 24 inches O.C.

Many elements of a wall frame call for specific insulation measures. Some call for unique treatment or changes in traditional framing. Others call for adding air sealing measures by stuffing insulation in connecting joints.

Headers

Insulate headers within the wall frame. Exterior walls must have supporting headers if they have windows or doors. Normally, these headers are a composite double 2 X members that are solid and thus able to allow conductive heat transfer. Instead of both members being

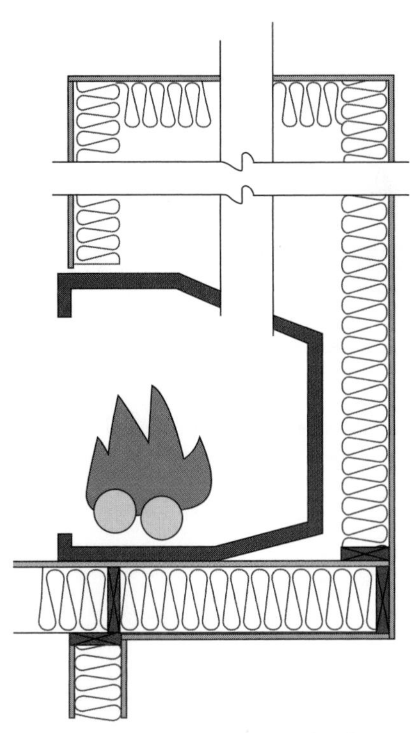

Courtesy of Southface/Earthcraft House

Figure 7-13 ■ A fireplace has historically been regarded as a heat source that generally results in a net energy loss for the home but adds a sense of emotional warmth. However, there are approaches to tightening up joints in a fireplace installation that improves its efficiency.

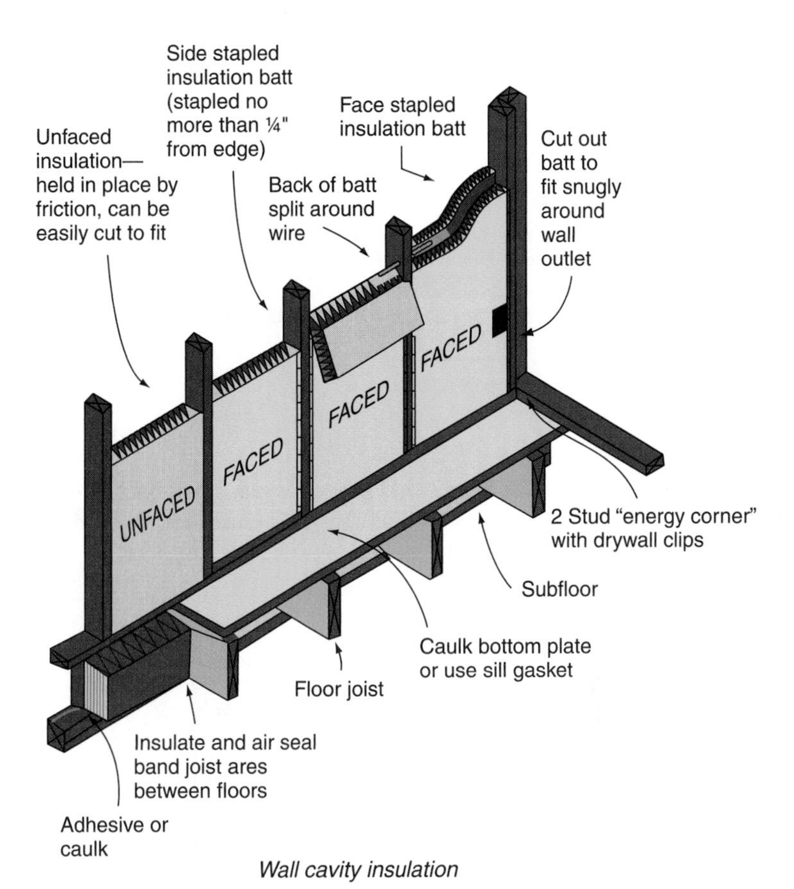

Courtesy of Southface/Earthcraft House

Figure 7-14 ■ Wall frame cavities must be completely filled with insulation to create a dead air space. Be careful with penetrations such as electrical outlets, and cut batt to fit snuggly. For wire within the cavity, split insulation around wire.

squeezed together, these headers should be built to accommodate ½ inch of rigid foam insulation between them. Additionally, the IRC allows the use of a boxed header with insulation batts within the space. Both conditions are illustrated in Figure 7-15 and provide the adequate structural support and an added bonus of thermal isolation.

Exterior band joist

Insulate the band joist that connects the floor joist to the exterior wall frame. This construction connection is illustrated in Figure 7-16. In the condition where an upper floor is significantly above grade, a band joist will define the end or edge of the floor framing joist. This floor framing will rest on about 2 inches of wall framing, allowing a band joist to rest on the edge of the top plate for a flush finish. Insulation is sometimes neglected on the inside of this band joist. Add insulation and seal any leak with caulk or thermal sealing measure.

Framed wall corners and intersecting walls

Walls intersect and must be tied together to form an effective structural connection. However, the joint can prevent air sealing measures unless plans are made to allow for insulating and sealing these joints. Insulated corners of two intersecting outside walls as shown in Figure 7-17 must be framed so that insulation is continuous in the exterior wall (by deleting the third stud). A two-stud corner with drywall clips is another method of achieving this seal.

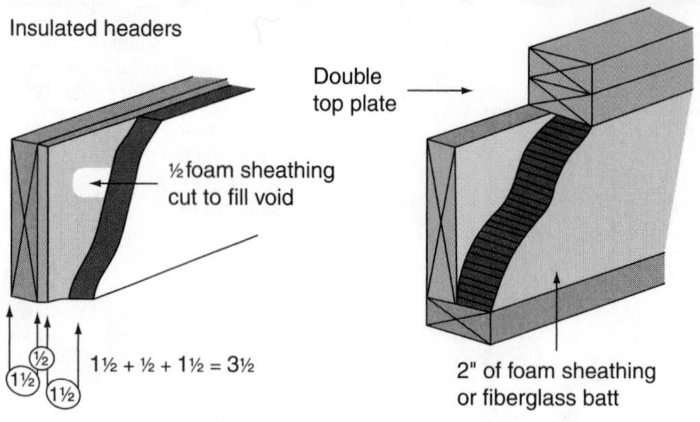

Courtesy of Southface/Earthcraft House

Figure 7-15 ■ Headers within a wall frame provide structural support but generally provide a conductive path for heat transfer because they are solid lumber. As an alternative, the use of insulated headers provides a thermal break between the lumber.

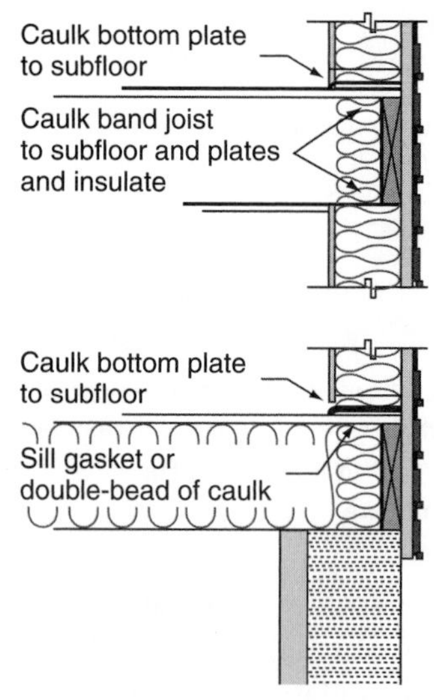

Courtesy of Southface/Earthcraft House

Figure 7-16 ■ As a wall frame transitions between floors, a band joist seals the end of floor joists. This band joist is normally solid lumber and can be a source of heat loss. The area within the cavity of the floor joist must be sealed against the band joist with insulation and caulking to retard heat flow through air leakage.

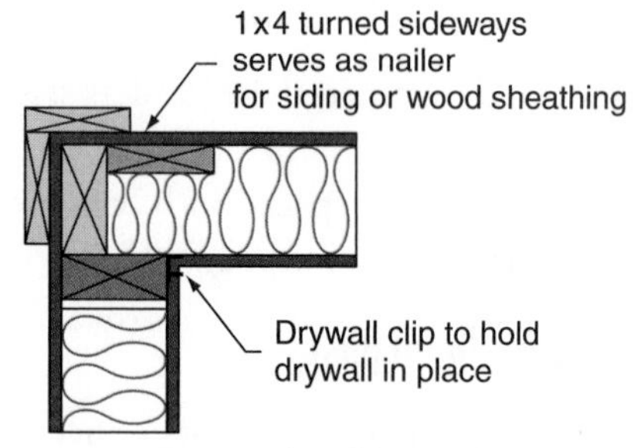

Courtesy of Southface/Earthcraft House

Figure 7-17 ■ Wall frames joined at corners normally have three studs. The Code allows for the elimination of the third stud with certain connections. In addition, this allows for ease in installing insulation around the corner.

Insulate intersecting walls of an interior wall and an outside wall with insulation that is continuous in the external wall. A ladder-type intersection, as illustrated in Figure 7-18, is one method of achieving this continuity and avoids the necessity of adding another stud, thus conserving materials as well.

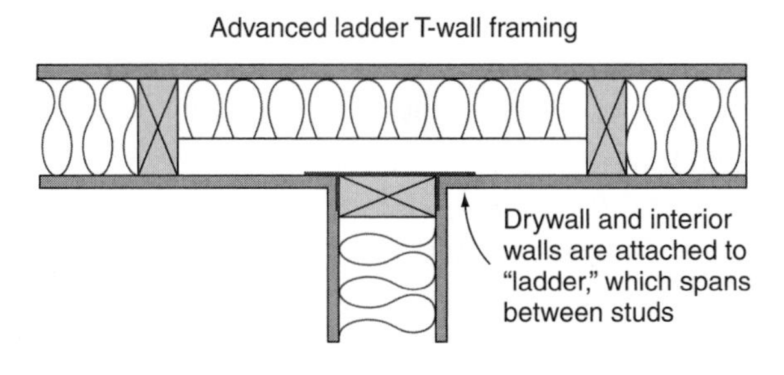

Advanced framing techniques add insulation.

Courtesy of Southface/Earthcraft House

Figure 7-18 ■ An interior wall may intersect an exterior wall and cause difficulty in installing insulation. An advance framing technique uses a *ladder* that spans between studs forming the structural connection and uses less building material while allowing for more thorough insulation in the exterior wall frame.

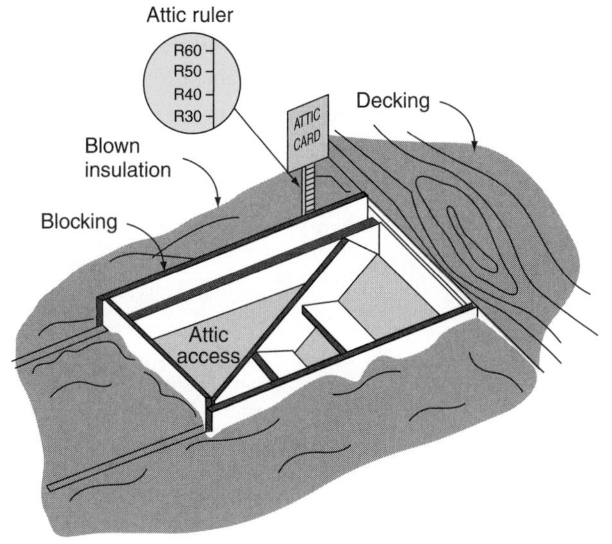

Courtesy of Southface/Earthcraft House

Figure 7-19 ■ Loose-fill or blown-in insulation can settle and may need to be replenished over time. This attic insulation must be measured to determine proper levels. An attic ruler adjacent to the insulation card provides for this measurement.

Insulated Wall Sheathing

Wall sheathing such as panelized siding or plywood serves as a wall brace to support the building frame, allowing the building to resist lateral forces such as wind or earthquakes. Sheathing material is available that not only has structural capacity, but carries the added component of being insulated. There are sheathing materials that have reflective surfaces that turn back radiant heat from the sun. A green home makes use of a building material or component that serves multiple purposes that include energy efficiency.

Loose-fill Insulation in Attic

Loose-fill attic insulation depth must be identified and marked. The thickness of blown-in or sprayed roof/ceiling insulation (fiberglass or cellulose) must be written in inches on a marker positioned every 300 square feet in the attic space. The markers should be connected to the roof system. They should indicate the minimum installed thickness of insulation. Each marker should be positioned toward the entrance to the attic as indicated in Figure 7-19.

Energy-efficient Truss Design

Energy-efficient design in trusses allows for the exterior wall top plate to have a raised top chord or plate to allow for increased insulation where it would normally be omitted. Commonly called a *raised heel roof (or energy heel)* truss, this type of truss allows full-depth insulation to be added. Note that this is designed by an engineer and is most commonly manufactured in controlled conditions. It is not the same as field framing. You cannot alter or change an engineered truss to create such a condition like you would with conventional framing. For this type of truss, exterior sheathing must extend above the top plate to the top of the heel on the truss, so you may need sheathing that is longer than the traditional 8-foot length. There are some cases where these types of trusses may not be appropriate because of structural considerations. In any case, required ventilation may still need to be provided.

Ceiling and Attic Insulation

Flat ceilings or ceilings with unconditioned attic space above should have complete coverage of attic insulation equal to or greater than R-38. Before installing attic decking, increase the joist height with an additional framing member on top. Be sure to secure properly and brace from lateral displacement with periodic blocking. Additionally, for ventilated attics such as those shown in Figure 7-20, be sure to install a baffle in the rafter to maintain ventilation from soffit vents.

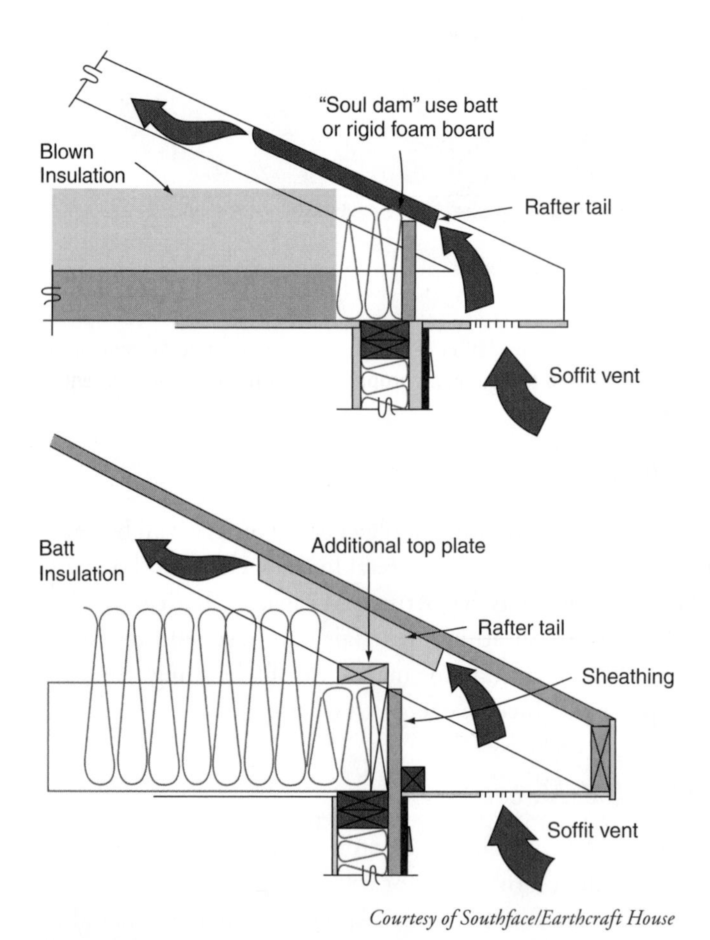

Courtesy of Southface/Earthcraft House

Figure 7-20 ■ Flat ceiling insulation in an attic area is critical to thermal envelope sealing.

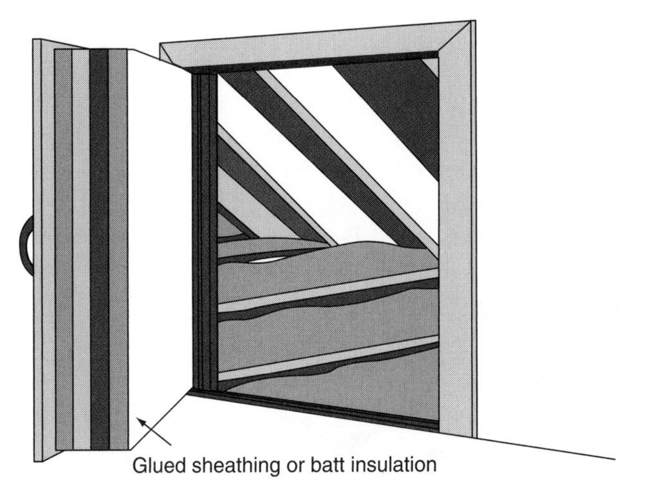

Courtesy of Southface/Earthcraft House

Figure 7-21 ■ Unheated attics with usable space for storage or maintenance are sources of heat loss through the access doorway. These doorways must be sealed and insulated as the adjacent ceiling or wall frames.

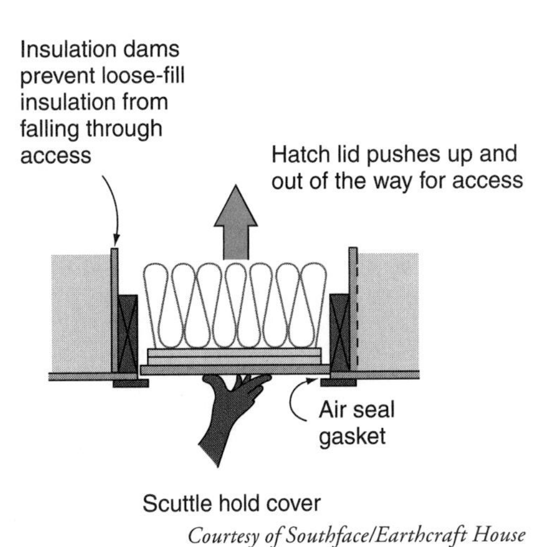

Courtesy of Southface/Earthcraft House

Figure 7-22 ■ Even if the access is a removable hatch, it should be insulated to retard heat loss.

Attic access doors should have an insulated box at the closure with at least an R-22. Scuttle holes that are located in conditioned space shall be insulated with batt insulation or rigid foam insulation to R-22 or greater.

Another source of energy loss is the attic access panels in homes with usable attics. As you can see in Figures 7-21 and 7-22, these must be installed with adequate insulation and proper air sealing measures. The side swing access door in Figure 7-21 would commonly serve the nonhabitable portion of an attic adjacent to a second floor. Although the storage is easily accessible, the air sealing measures are usually ignored. Adding a layer of insulation to the scuttle access as seen in Figure 7-22 is not sufficient. A gasket to seal the air leaks is essential to maintain energy efficiency. Vaulted ceilings should be insulated to an R-38 if possible. It may be difficult with the narrow width as the scissor truss ends join together. Remember, the value of the insulation is diminished if the insulation is bunched up or compressed. The R value is realized when the total dead air space is developed. If the space is 6 to 8 inches, forcing in R-38 batt insulation that is normally 10 inches thick will not equal the designed thermal performance. A solution would be to install a scissor-type or an energy heel truss that would accommodate the installed insulation.

Ceiling radiant barrier

Solar energy is absorbed or reflected depending on material properties. Solar energy can be reflected away from the home with a radiant barrier acting much like a mirror. Ceiling radiant heat barrier like the one shown in Figure 7-23 should be installed over a large portion if not all of the vented attic space. The radiant barrier must have a reflective surface facing toward the conditioned air space, a minimum of 1 inch vented air space, and have an emissivity rating of 0.05 or less.

Doors and Windows

The most common means of sealing window and door leaks is by weather stripping. When the window or door is installed, ensure a tight fit and an adequate seal around the frames. In a

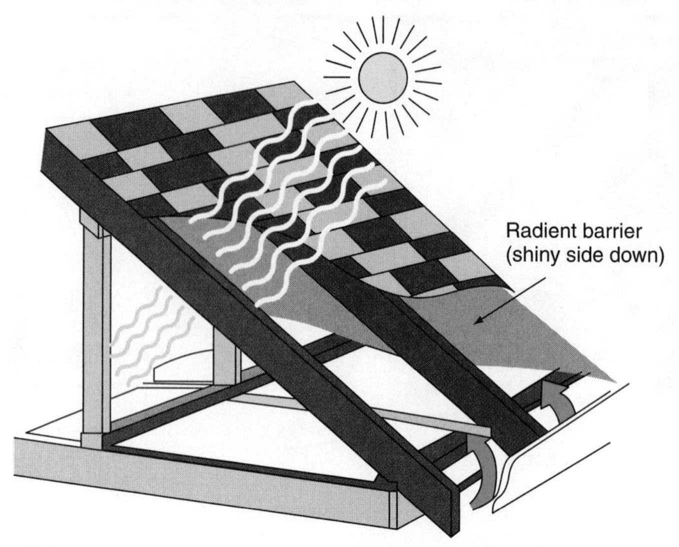

Courtesy of Southface/Earthcraft House

Figure 7-23 ■ Radiant barriers reflect solar radiant heat and retards heat gain in a habitable space.

wood frame wall, there are trimmer studs, cripple studs, and headers that would normally be the connection points for the window or door. Sealing in these cases as illustrated in Figure 7-24 involves caulking and insulation. For drywall, connection to the surface of these framing elements should include more than just nails. Run a bead of caulking before nailing to improve the seal when nailed. Rough window and door openings should be sealed with nonexpanding spray foam sealant or other suitable sealant. Fiberglass insulation alone is not enough. Thresholds for doors should be sealed directly to the floor surface to fill the gap.

Window and door warranty and manufacturer's installation instructions should be reviewed before selecting a sealant. Note that expanding-type sealant can damage window frames and void warranties.

Window and door thermal efficiency must exceed that required by Chapter 11 in the IRC. A μ value that is 40% lower than required is desirable. The required μ value varies depending on climate zone but varies between 1.2 and 0.35 (remember, the μ value is the inverse of the R value). Windows and doors represent fenestration. These are typically more prone to heat loss because it is more difficult to insulate. To ensure that windows and doors meet energy performance standards look for a certification from the National Fenestration Rating Council (NFRC) label as illustrated in Figure 7-25 with appropriate μ values for your climate zone.

- NRFC windows: Windows should be energy rated by the NFRC and have a U factor of 0.40 or less.
- Low emissive windows: Windows should have low emissive-type glazing. All windows should have two glazing layers and the inner surface of one layer should contain a low emissive coating.
- Gas-filled windows: High-efficiency windows are desirable. Inert gas-filled double glazing units (argon gas) are highly efficient. Double paned windows have an insulating gas, such as argon or krypton, between the two panes.

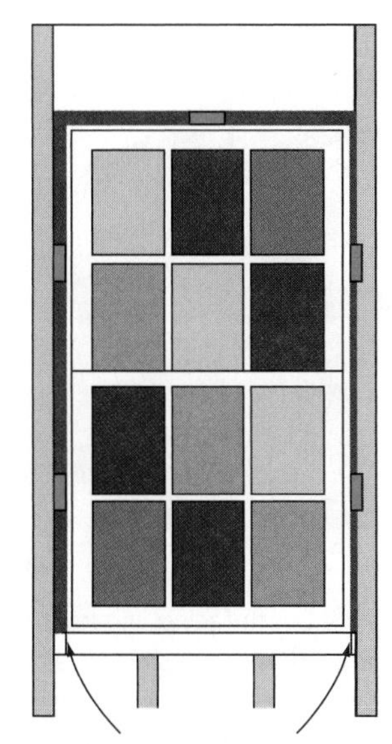

Courtesy of Southface/Earthcraft House

Figure 7-24 ■ Windows and doors are installed within a wall frame. The joint between the window and wall frame must have a seal. It could include a nonexpanding spray foam sealant or another suitable sealant that does not void the window or door manufacturer's warranty.

Courtesy of NFRC

Figure 7-25 ■ A typical National Fenestration Ratings Council certification label.

Energy Efficiency and Windows

Windows that have a low SHGC (max 0.40) are desirable in climates with solar heat gain potential. Windows facing east, west, and south shall have a SHGC of 0.4 or less. A properly designed roof overhang will block the summer sun but allow the winter sun to warm the same space. A roof overhang should be installed above windows on a south-facing wall. Windows should have an overhang with dimensions based on where the top and bottom of the window are located in relationship to the overhang in order to provide shading. In most cases, a minimum 1½-foot overhang is sufficient to protect against solar heat gain.

Solar shade screens are available for windows facing east and west. The screening will have a shading coefficient of 0.7 or greater. Shade screening should be installed on the exterior of window glazing. Exterior shade screens must reduce solar heat gain through windows by 70%.

Solar Design

Passive solar design and installations should be capable of reducing heating loads by 25% or greater while not increasing cooling loads by more than 10%. These passive designs include glazing with southern exposure with a thermal mass storage facility such as concrete or masonry wall or floor. The passive solar heat gain could be made by one of several means involving solar heat collection through glazing. The most common method is through direct gain.

Direct gain is accomplished by collecting solar heat directly from the sun shining into a building, most commonly through south-facing windows. The solar heat can be collected and maintained in a thermal mass device for delayed use when the sun goes down. A thermal mass

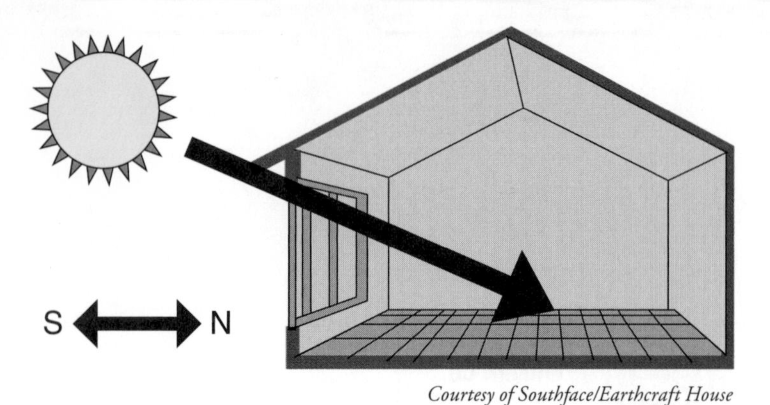

Courtesy of Southface/Earthcraft House

Figure 7-26 ■ Appropriate placement of windows based on the solar orientation of the home will allow heat gain in winter and avoid heat gain in summer.

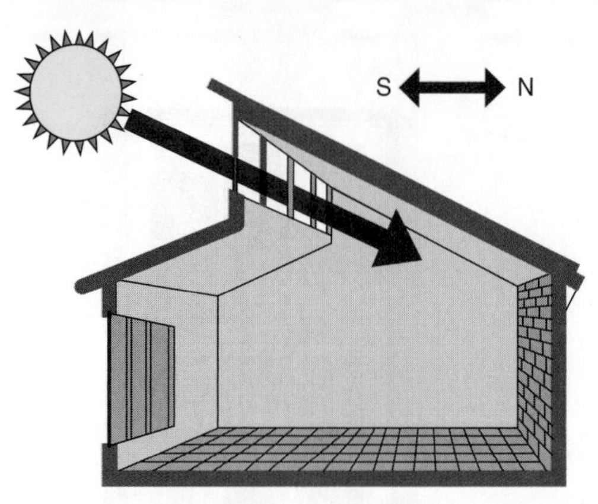

Courtesy of Southface/Earthcraft House

Figure 7-27 ■ South-facing windows in the northern hemisphere will net a positive heat gain from passive solar energy. Clerestory windows like this allow for deep penetration of solar radiation onto interior walls.

can be any material that can absorb and store heat. Materials such as concrete or masonry are the most common choices for thermal mass in a passive solar home. These materials absorb solar heat and release it slowly when heat is more appropriately needed.

Active solar heat gain involves the use of mechanical means to transfer the collected solar heat to another part of the building. Active solar heat gain collects heat from the sun and conveys it to other spaces within the building. Solar heat accumulated within a sun porch can circulate naturally through convection or mechanically with the use of fans to other rooms within the building. In all cases, there are several basic considerations for solar heat collection:

- The building should be oriented toward the sun on an east-west axis.
- The building's south-facing wall must receive sunlight during the daytime in the heating season.
- Interior spaces that need the most natural lighting should be adjacent to this south-facing wall.
- The building must have an open floor plan that allows passive solar heating and natural lighting to work.

The gain of solar heat in the winter must be mitigated with control of heat gain in the summer months. This is normally done with solar shading that extends over the top of the window—the length necessary to prevent excessive heat gain. This is illustrated in Figure 7-26. Make use of solar heat in the winter or heating season by allowing the sun in as illustrated in Figure 7-27.

The Department of Energy asserts five elements that constitute a complete passive solar home design:

1. *Aperture (collector)*: A large glass (window glazing) area through which sunlight enters the building. Typically, the aperture should face within 30 degrees of true south and should not be shaded by other buildings or trees from 9 A.M. to 3 P.M. each day during the heating season.

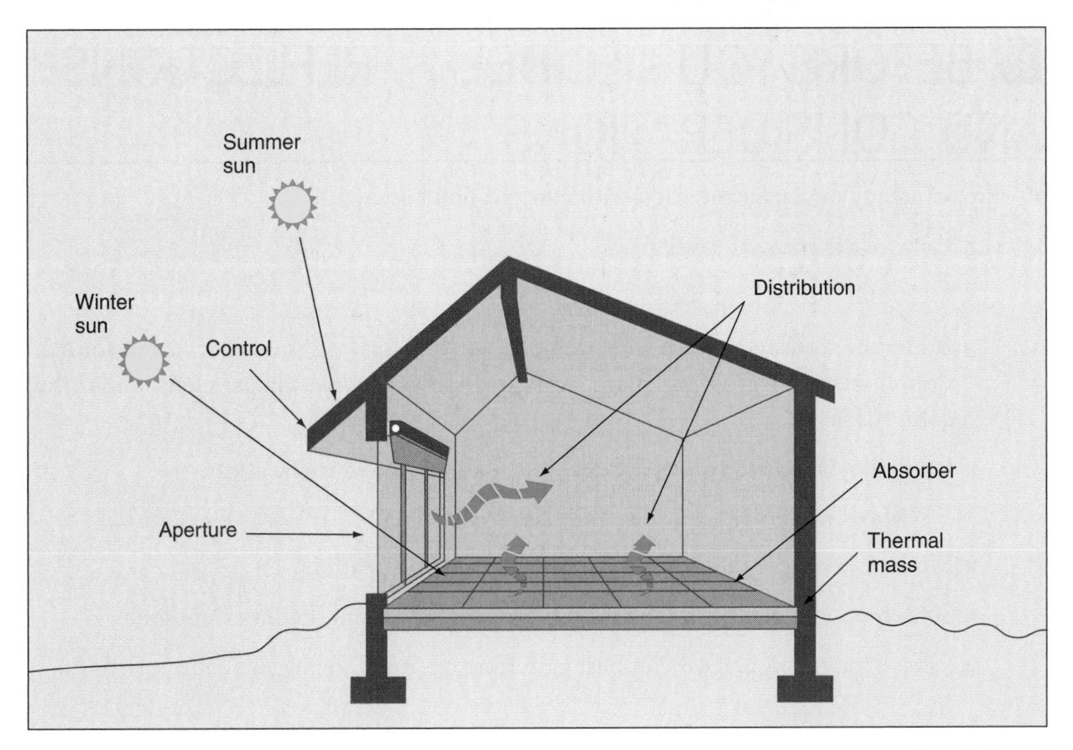

Lynn Underwood

Figure 7-28 ■ A passive solar home uses south-facing windows with the proper roof overhang to control heat gain in the heating season and avoid heat gain in the cooling season.

2. *Absorber*: A hard, darkened surface of the storage element. This surface—which could be that of a masonry wall, floor, or partition (phase change material), or that of a water container—sits in the direct path of sunlight. Sunlight hits the surface and is absorbed as heat.

3. *Thermal mass*: Materials that retain or store the heat produced by sunlight. The difference between the absorber and thermal mass, although they often form the same wall or floor, is that the absorber is an exposed surface, whereas thermal mass is the material below or behind that surface.

4. *Distribution*: A method by which solar heat circulates from the collection and storage points to different areas of the house. A strictly passive design will use conduction, convection, and radiation exclusively. In some applications, however, fans, ducts, and blowers may help with the distribution of heat through the house.

5. *Control of heat gain*: Roof overhangs can be used to shade the aperture area during summer months. Other elements that control under- or overheating include electronic sensing devices, such as a differential thermostat that signals a fan to turn on, operable vents and dampers that allow or restrict heat flow.[7]

The value of passive solar heat collection systems such as the one illustrated in Figure 7-28 includes minimal maintenance, no regular operating costs, and no by-products that affect the interior or exterior environment. Although this system is unlikely to provide all of your heating needs, it can substantially reduce the need for normal energy usage that would be required without such a system.

[7] *U.S. Department of Energy, Energy Efficiency and Renewable Energy (EERE), http://www.eere.energy.gov/*

BEFORE YOU DECIDE . . . REFLECTIONS AND CONSIDERATIONS

✔ Proper energy management starts with conservation basics.

- Control the thermal envelope.
- Meet and then exceed the Energy Code.
- Build the tightness of the construction with air sealing measures in the foundation, floors, walls, roof, fireplace, and in windows, doors, exhaust ducts, and utility connections.
- Insulation is indirectly proportional to the ability to reduce heat loss.
    - More insulation equals less heat loss (or heat gain).
- Modern developments in framing allow for superinsulation.
- Solar heat gain can be mitigated with proper glazing and shading techniques.
- Passive solar design allows for heat gain from a natural, renewable source: the sun.

FOR MORE INFORMATION

Energy Star
http://www.energystar.gov/

U.S. Department of Energy, Energy Efficiency and Renewable Energy
http://www.eere.energy.gov/

Chapter 8

LIGHTING AND APPLIANCES

LIGHTING

Lighting is a very important aspect of a home. For architects and designers, interior lighting is paramount for creating a superior design for living. There is a need for good quality—natural light that does not create glare and good artificial light at night. Important in both cases is light that does not add to unwanted heat.

Lighting establishes the mood for each room in a home during different times. Natural light affects our health, producing vitamin D and regulating our circadian rhythms. Sunlight enters the home in one of several ways. It can enter by reflection from ground surfaces, adjacent buildings, and other similar objects. Light also comes from direct sun exposure from a skylight or window facing the sun. The most common way that light is brought into a house is with a reflected component. Sunlight enters a room through a window and strikes the surrounding walls, ceiling, and floor surfaces, which serve as reflectors. If the reflected light is excessive, it results in glare. This is important for daytime illumination. During the evening there is little or no natural light in our homes. Artificial lighting allows us to extend our productive day and enjoy our lives in many other ways. However, lighting comes at a cost. Lights that draw electricity account for 19% of all energy consumed in the United States according to the Department of Energy. Consequently, the more natural light gained, the greater the reduction in artificial lighting that can be realized. Correspondingly, energy-efficient artificial lighting rounds out the equation for necessary illumination when natural light is not available.

Daylighting

Daylighting is an effort to bring sunlight into a building. It uses the sun's natural light to illuminate the interior of buildings. Efficiency is smart, but reducing the need for energy has better long-term consequences. A successful daylighting process brings indirect outdoor lighting into a building while avoiding the associated heat gain or loss that normally comes with fenestration. This type of process reduces the need for electrical lighting and is a fraction of the cost.

Although it is possible to mitigate heat gain or loss, reducing the need for electric lighting and cooling from the start is better. Cool daylighting does both. Daylighting can be achieved without having to increase construction costs and is especially cost-effective in new construction. Electric lights in buildings create heat that must be dissipated to allow for a comfortable living environment. Less electric lighting means less heat gain, which means less heat to remove with air conditioning, thus using less energy. Natural lighting that generates almost no heat can be provided. By doing so, it is now possible to downsize cooling systems. The associated cost reduction in selecting a smaller unit helps cover the cost for the daylighting installation. To avoid heat gain while adding light, properly designed daylighting screens out most of the sun's heat while still providing indirect light.

Any light source, if intense, can produce glare. The traditional solution to this common occurrence is drapery or directional blinds. There are ways to passively avoid glare in your home. You can choose where to place certain windows, considering the angle of the sun throughout the year. You can select shading devices, such as draperies or overhang on the roof or deck. Low-transmittance glass is available to reduce or block direct sunlight and significantly reduce glare. Directing light to enter a room from two different angles by placing windows on different walls can reduce glare.

Natural light reduces the necessary amount of electric lighting, but accommodation is necessary for nighttime use. Remaining heat created by this necessary light can be reduced by selecting the proper fixtures or resolved by installing energy-efficient equipment. Clear fenestration lets too much light inside. The sun provides 5,000 to 10,000 foot-candles of light, depending on Earth-sun position, weather conditions, season, and location of the site. Necessary interior lighting for a home would vary between 50 and 100 foot-candles depending on the uses of the room or space. Even in poor weather conditions, at least 5,000 foot-candles are provided by the sun, far more than are needed for the interior of your home.[1] More than that would be perceived as glare and less would be too dark to use. Daylighting lets in natural light that balances overhead electric lighting while curtailing glare. The following is a summary of the benefits of daylighting:

- Even, more comfortable lighting, which can reduce daytime eye fatigue
- Improved life-cycle cost for initial installation and maintenance
- Increased use of room or space while decreasing energy costs
- Improved quality of interior environment with better light
- Reduced energy usage
- Decreased environmental damage

Skylights

Another means of bringing in natural lighting is with a skylight as illustrated in Figure 8-1. This method of daylighting best serves large interior spaces that are ordinarily without windows. Usually, if a room or space is a distance from an exterior wall or door, a skylight serves it well. A type of window that is very successful for solar collection is a clerestory window. These are windows set in a row at the top of the wall that usually is above an adjacent roof line

[1] *Emily Rabin (2006, April 18), "Harnessing Daylight for Energy Savings," GreenBiz.com, http://www.greenbiz.com/*

Courtesy of iStock Photo

Figure 8-1 ■ Skylights are a proven method for bringing natural daylight into interior spaces that must otherwise be artificially illuminated.

that intersects below the windows. They are effective for tall ceilings and can serve as thermal collectors of solar radiation.

Windows usually constitute a fourth to a half of a wall's space. For traditional skylights, use glazing material that reduces glare. Clear glazing used in skylights should employ some sort of a diffuser at the bottom of the skylight shaft to create indirect natural lighting. Double glazing is recommended as a minimum for skylights. For skylights with a shaft that opens into the habitable room, install an insulating panel to prevent heat loss in winter months unless the skylight is insulated. Energy loss is inevitable. Warm air rises and tends to migrate toward skylights. Even with double glazing, there will be loss of heat. Adding solar tubes, skylights with reflective shafts, can bring light farther into a home with limited heat gain or loss.

Light Shelves

A light shelf is a passive architectural technique that invites sunshine farther inside a building. Light shelves can replace the need for lighting near the perimeter of the building. With the appropriate overhang, exterior light shelves may also function as sunshades. An estimate

Courtesy of Ann Arbor District Library

Figure 8-2 ■ Light shelves use horizontal light-reflective surfaces to divert light farther into interior spaces. This limits the need for artificial lighting. Although light shelves are generally installed in commercial applications, large homes would also benefit from the same principle.

is that light shelves can penetrate 2½ times the distance from the floor and the top of the window. As an example, a home that contains a light shelf that is 7 feet high would produce a projection of over 17 feet. A light shelf, such as the one illustrated in Figure 8-2 is either a fixed or an adjustable device that reflects or shades light that enters through a window, including task lighting and uplighting. A light shelf using uplighting can reduce glare caused by direct lighting applications.

Roof Monitors

A roof monitor is a flat roof section raised above the adjacent roof with glazing installed on all sides, like a small penthouse, such as the one depicted in Figure 8-3a. A skylight, as illustrated in Figure 8-3b, with a tall curb would be a rudimentary example of a roof monitor. Other more exotic types include angled glazing that approximates a right angle to the sunlight. As with skylights, these tend to collect warm air because warm air rises. With very good operable windows, roof monitors can serve in natural ventilation schemes as well.

Because of the large surface area for glazing, the roof monitor is not as efficient as a clerestory design. The multiple angles possible for the glazing can allow for design reflection but also can create shading problems.

(a)

(b)

Courtesy of iStock Photo

Figure 8-3 ■ Roof monitors provide uniform lighting to deep interior spaces and eliminate glare.

Design Elements

There are several elements of design that also add to the efficiency or compensate for the cost of illuminating your home in an appropriate manner. There are many avenues toward environmental friendliness that carry a price. Adding fenestration such as a window decreases the effectiveness of the thermal envelope. In addition the heat gain realized in the summer months exacerbates the problem. Design compensation could include positioning windows at the cardinal points around the house to collect winter heat and avoid summer heat. A few other ideas follow:

- Use fewer windows and doors on the north side. Those used should be high efficiency, low μ value, low e value.
- Clerestory windows can allow for illumination of large interior rooms. The use of tubular skylights, light shelves, and roof monitors all increase natural daylighting while controlling heat loss through air leakage. The most appropriate direction for the clerestory to face is south or north, not east or west, because at least twice-a-day direct sunlight tends to cause glare.
- Tall windows bring light farther into a room than short, wide windows.
- In hot climates use fewer windows on the west side and install an overhang to shade the summer sun on windows that face south.
- Design locations for windows to illuminate most of the space within rooms through indirect lighting.
- Install windows to allow for cross-ventilation of fresh air. An operable skylight or window high on a wall can let hot air out in summer and gain heat in winter.
- Create a balance in daylighting by adding fenestration on the opposing side. This will diffuse the light and decrease glare.
- Decorate to complement the daylighting. Use lighter colors for floor covering, walls, and ceiling to allow for further reflectivity. This design element helps with artificial lighting as well.
- Use an *open floor plan* in the design stage to allow for more use of natural lighting.
- Large overhangs up to 4 feet will provide shade in the summer yet allow light inside a building during winter months when the sun in lower in the sky.

Figure 8-4 illustrates how an overhang properly shades solar heat gain in the summer while allowing solar heat gain in the winter. Dividing the height of a wall that needs shading in the summer by a factor of 4 is optimum for a projection for windows on southern walls.

Integrated Building and Construction Solutions (IBACOS) has created a guide for high-performance lighting (HPL). It is a useful tool and has seven sections that will help you become familiar with HPL. These include:

- Understanding High-Performance Lighting
- Using This High-Performance Lighting Guide
- IESNA Guidelines
- Room-by-Room Designs

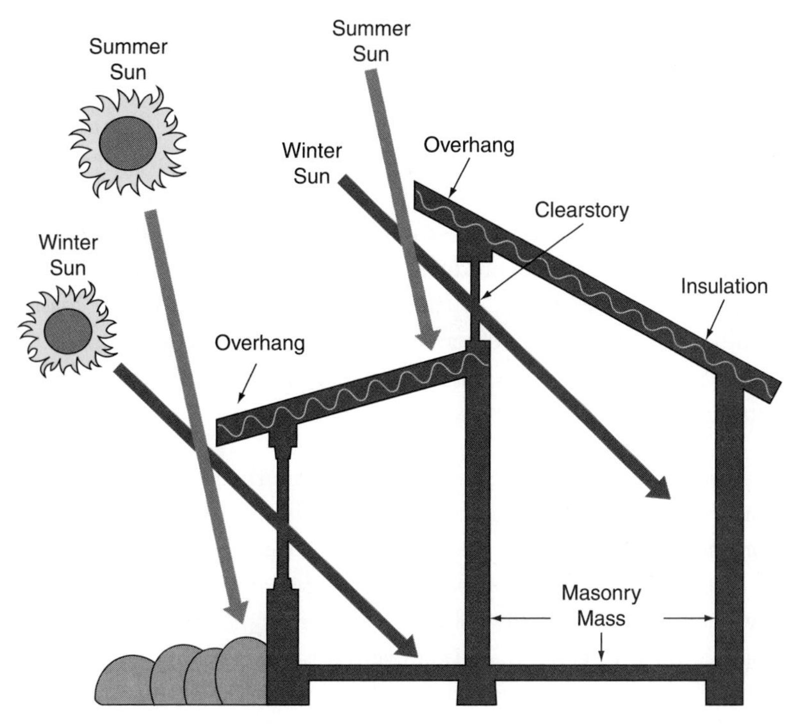

Delmar/Cengage Learning

Figure 8-4 ■ With the precise orientation based on latitude, a south-facing window can gain passive solar heat in the winter and shade it from entering with the proper roof overhang in the summer.

- Specifications for Fixtures
- Specifications for Lamps and Bulbs
- More Resources[2]

Energy-Efficient Lighting and Bulb Selection

When considering artificial lighting, evaluate the need based on the room usage. For example, a kitchen counter space should be well lit because culinary chores require adequate illumination, whereas a dining area in the same kitchen can have a reduced lighting level. Thoughtful consideration is needed to provide the right type and amount of light for each area of your home. Modern, efficient lighting that can reduce electrical consumption for lighting is available and lighting design can actually increase a fixture's effectiveness.

Legislation

The Energy Policy Act of 2005 is the largest overhaul of the national energy policy since the Energy Policy Act of 1992 and directly affects lighting. This policy began the phased elimination of certain kinds of incandescent lighting and established standards of efficiency for some commonly used lamps. It even banned some types of lamps. Because of this policy, there are new ways of shopping for lighting. Lighting is no longer determined by watts. Now, lumen output helps the consumer compare various products. In addition, the color cast by certain lights (notably fluorescent) is evaluated through the use of a color rendering index (CRI).

[2]*IBACOS, http://www.ibacos.com/*

Fluorescent Fixtures

Four times as much energy is converted to visible light in fluorescent lamps in comparison to incandescent bulbs. Clearly, energy efficiency is a strong attribute of fluorescent lights. However, one difficulty with fluorescent lamps is the color rendering of objects illuminated by this type of light. A fluorescent lamp or bulb is one that sends electricity into an ionized gas such as mercury vapor or neon that changes into a state of plasma and discharges short-wave ultraviolet light causing fluorescence, producing visible light that is generally toward the shade of violet. Color rendering refers to the appearance that a light gives to the color of an object. A CRI rates a bulb on a scale between 0% and 100% and indicates the accuracy that a particular bulb provides. A higher CRI will result in a better bulb color rendering ability. The CRI is a crucial part in the selection of a fluorescent fixture and lamp.

Fluorescent lamps generally require a ballast to regulate the flow of electricity through the lamp. In common tube fixtures the ballast is located outside the lamp but within the fixture and is most commonly a rectangular or cylindrical device. The ballast provides the starting energy and limits the current to the proper value for the lamp. The fluorescent lamp tries to produce light without heat. Electrons collide with mercury atoms that produce a discharge of ultraviolet light. This ultraviolet light is converted to light that we can see when it passes through a layer of white phosphor powders inside the lamp's glass envelope. The fluorescent lamp produces this visible light with very little thermal energy as a by-product.

From a green perspective, the fluorescent lamp brings certain savings and environmental advantages but at a cost. The use of mercury vapor is essential to create the fluorescence effect and visible light. There can be a detrimental impact on the environment if the lamp is disposed of improperly. However, there are low mercury bulbs available and more recycling for these lamps than before and the trade-off seems appropriate given the offset of savings in energy and life-cycle alternatives. In addition the introduction of mercury vapors into the habitable living area, if the bulbs were inadvertently broken, would tarnish the indoor air quality.

Compact Fluorescent Light Bulbs

Using compact fluorescent light (CFL) bulbs instead of incandescent bulbs can significantly reduce your energy bill. These bulbs use about one-fourth the energy of the incandescent bulbs used in homes today compared to a similar luminary output. Several energy providers claim that replacing one 60-watt incandescent bulb with a CFL bulb can save $54 over the life of the bulb. This efficiency also translates to reduced heat output, which normally reduces cooling loads and furthers savings. There are several bulb types to consider, such as straight tube, twisted tube, and folded tube. Most manufacturers claim a life that is 8 to 10 times that of the incandescent bulb—some as much as 10,000 hours. CFLs are available in 15- and 20-watt bulbs. Some consumers report that a lamp bulb must be turned on for at least 3 or 4 hours at a time. Otherwise, they will burn out in less than a year. Advances, including the electronic ballast, have allowed these lamps to be small enough to be viable for everyday use in most lamp bases with a screw base. Cold cathode compact fluorescent bulbs use a cold cathode (instead of hot) to release electrons into the compact fluorescent tube. This technology, though using more electricity, allows for instantaneous illumination with the use of a switch. CFLs contain mercury that must be disposed of in a proper manner when the bulb's use has ended.

Courtesy of iStock Photo

Figure 8-5 ■ LEDs are very small light bulbs that fit easily into a small electrical fitting. They do not have a traditional filament that burns out.

Light-Emitting Diodes

Light-emitting diodes (LEDs), as illustrated in Figure 8-5, represent a simple solution and immediate energy savings. The LED is generally not hot to the touch like an incandescent bulb although it does produce heat. LEDs last longer than traditional incandescent bulbs. Incandescent type bulbs produce light by heating a small filament of tungsten to about 2,500°C. LEDs are very small light bulbs that fit easily into a small electrical fitting. They also do not have a traditional filament that burns out. They produce light from the motion caused by electrons moving through a semiconductor material such as silicon.

Torchieres

Torchieres are portable light fixtures that project toward the ceiling, allowing for indirect lighting to reflect and illuminate the walls and floor area. They are usually on a tall stand of wood or metal and have a shade that directs light upward. This type of lighting works best where the ceiling is not much higher than 8 or 9 feet. Torchieres are satisfying, popular lights, but the bulb is generally a high-wattage halogen lamp. Although it gives a soft white light, those with halogen lamps operate at very high temperatures—near 1,000°F—and can pose a fire hazard if the manufacturer's directions are not followed.

Now a newer innovation is a compact fluorescent torchiere. This compact version combines the design feature of ceiling-illuminated indirect light with the energy efficiency of a compact fluorescent lamp that draws around 60 watts. These also operate at a cooler temperature—near 100°F—posing less fire hazard. A standard appliance has one or two compact fluorescent lamps.

Cove Lighting

Cove lighting is background lighting. It is a form of indirect lighting that emanates from ledges, valances, or horizontal recesses on the upper wall or in the ceiling of a room. This provides an even, indirect lighting of floor and wall spaces. This type of lighting is characterized by low-wattage fluorescent lamps or incandescent bulbs. These lights conserve energy and have a longer operational life. Cove lighting can provide limited light for a specific subject of illumination, or it can add accent by providing a specific color.

This type of lighting often provides a soft, warm feel to an architectural feature. Light from this type of fixture is normally directed upward, reflected off the ceiling or upper portion of a wall. This reflected light is even and mild without glare. Cove lighting creates a relaxed, ambient light that promotes relaxation. The light fixtures used for cove lighting can be low-voltage fluorescent or incandescent bulbs. These bulbs provide an adequate amount of light. To create a certain mood, these types of lights can be controlled with a dimmer switch so they can emit a small amount of light. Incandescent and fluorescent lamps are both capable of providing colored light if desired. Fluorescent lamps use electricity efficiently in providing a given amount of light. Cove lighting can be used in many locations in your green home such as bathrooms, bedrooms, dining rooms, game rooms, hallways, kitchens, libraries, and living rooms. To increase its effect, the top of the light source should be around 12 inches below the ceiling and positioned to allow the light to spread over the ceiling. A dramatic effect can be achieved by adding color to the wall that contrasts with the light source.

Demand Responsive Lighting, Dimming, and Occupancy Sensors

Habitable rooms in a home require natural or artificial light. Most designs provide for both types of light based on use of each as the occasion dictates. Making a conscious effort to dim the light in a room is a very effective way of reducing energy usage. There are two general types of dimming fixtures. One is a varying voltage that goes to the ballast with a transformer or by use of electronic ballasts. Another is by use of dimmable ballasts. A fluorescent light with a daylighting type of ballast will allow the lamp fixture to vary the output based on the current needs of the room or space and on the ambient conditions. In other words, this daylighting ballast uses only the energy needed to bring the necessary light to the room based on the ambient conditions.

In rooms without natural lighting, an occupancy sensor can switch the light on when someone enters a room and off when it no longer senses motion within it. An interior bathroom is a room that would benefit from a technique like this one. A similar mechanism that saves energy is a light with a timing control. This is one that turns on the light for a set amount of time, usually in minutes. These timers can turn a light on (and off) at predetermined times.

Outdoor Lighting Controls

As with interior lighting, outdoor lighting is useful and even provides a safety and security function. Because of this, demand responsive lighting or sensors are especially appropriate for outdoor conditions. The cost of energy is significantly reduced when lighting is limited to demand conditions such as arriving guests. Motion or photo-sensors are capable of activating this type of outdoor lighting. When you are away from home for an extended time, a timed light can feign presence and is especially beneficial. In either case, energy demand is reduced through the selection of this type of outdoor lighting.

Design Elements

As with natural lighting, there are design elements that you can make in your home that directly affect energy usage. These range from the style of fixtures that perform a specialty task to sensor-controlled lighting. When deciding on the aspects of lighting desired for your home, various general categories of lighting should be considered. Ambient lighting is the available lighting in a room or space such as the general illumination for doing daily activities. Task lighting is the necessary lighting that allows us to complete specific household tasks accurately. Accent lighting illuminates specific features, such as a wall, to allow for blending with ceilings and windows.

Avoiding Unnecessary Use

The misuse of lights in a home is wasteful. It is estimated that 20% of our electric bill is spent on lighting. The behavior of the occupant has a great deal to do with the use or misuse of lighting. If possible, avoid the use of unnecessary lighting fixtures. Turning lights off that are not necessary is a very meaningful way to make a positive change. If we had meters on each appliance like we do on our electrical service equipment, we would be alarmed with the rapid pace of the electricity being used to generate light that otherwise would be wasted. If no one is in the room, turn off the lights. You will save money and energy and reduce environmental impact.

APPLIANCES

Your home consumes energy in the form of electricity in a variety of ways. A comparison of the most common appliances and their respective energy demands is depicted in Table 8-1. The largest single demand is in the form of electrical heating and cooling, followed closely by water heating and cooking equipment. After these heavy demands, the largest is from appliances, mainly those in the kitchen. Based on the demand of the traditional electrical installation, it is estimated that up to one-fourth of your total electric bill is accumulated from within the kitchen. Your green home has great potential to reduce electrical demand. Estimate energy demand of an appliance by looking at the nameplate on each appliance and determining its demand in watts. Wattage is simply volts multiplied by amps (the other name for watts is volt-amperes). This is not the actual demand; it is the maximum demand for the unit. Because the appliance is not always running, it does not meet the maximum demand. Focusing on the appliances with a major or significant demand, there are appliances you can shop for that will add energy efficiency to your new green home.

Energy-Efficient Refrigerator

Compared to other appliances in the kitchen, the refrigerator makes a significant demand on your electricity. Following are several design features that will enhance its energy efficiency:

- Energy Star label
- Energy Guide label
- Freezer on top or bottom
- Manual defrost cycle
- Refrigerator door alarm (if left open)

Energy Efficiency Labeling

Government evaluation of products can help you make an informed decision on what appliances to buy. The U.S. Environmental Protection Agency places an Energy Star label on refrigerators that are at least 15% more efficient than the federal minimum or 40% less than

■ Table 8-1

The Comparative Monthly Usage and Costs of Various Appliances in a Kitchen				
Appliance	**Typical Wattage**	**Average Hours Used**	**Average Monthly kWh**	**Typical Monthly Cost**
Coffee Maker	900	13	12	$1.61
Compactor (Trash)	400	10	4	$0.54
Deep Fat Fryer	1,500	3	4	$0.54
Dishwasher (With Dry Cycle)	1,000	20	10	$1.34
Dishwasher (Without Dry Cycle)	200	20	4	$0.54
Freezer—Upright/Chest 17 ft^3	600	720	173	$23.23
Freezer—Upright/Chest 17 ft^3—Frost-free	600	720	216	$29.01
Lighting—100 Watt	100	240	24	$3.22
Lighting—75 Watt	75	240	18	$2.42
Lighting—Comp Fluorescent—18 Watt	18	240	4	$0.54
Lighting—Comp Fluorescent—23 Watt	23	240	6	$0.81
Lighting—Fluorescent 2 Bulb	100	240	24	$3.22
Lighting—Halogen	90	240	22	$2.95
Microwave Oven	1,500	10	15	$2.01
Mixer—Stand	300	20	6	$0.81
Oven	5,000	10	25	$3.36
Range—Large Surface Unit	2,400	10	24	$3.22
Range—Small Surface Unit	1,200	10	12	$1.61
Refrigerator—14 ft^3	226	720	65	$8.73
Refrigerator—14 ft^3—Frost-free	383	720	91	$12.22
Refrigerator—17 ft^3—Frost-free	463	720	110	$14.77
Refrigerator—19 ft^3—Frost-free	509	720	121	$16.25
Refrigerator—21 ft^3—Frost-free	572	720	136	$18.26
Refrigerator—Freezer over 21 ft^3 Side by Side	783	720	186	$24.98
Refrigerator—Freezer 24 ft^3—Frost-free	653	720	155	$20.82
Refrigerator—Freezer 25 ft^3—Side by Side	841	720	200	$26.86
Slow Cooker	200	40	8	$1.07
Toaster	1,000	3	3	$0.40
Toaster Oven	1,500	25	10	$1.34

Note: Excerpts from the table are reused with permission from the Public Service of New Hampshire.

conventional models sold in 2001. These labels will help you understand how efficient an appliance is when compared to the average. The yellow Energy Guide label is a guide to help the purchaser with a buying choice. All major home appliances must meet the Appliance Standards Program set by the U.S. Department of Energy (DOE). Manufacturers must prove the energy use for their appliances using standard test procedures developed by the DOE. These results are printed on a yellow Energy Guide label, which manufacturers are required to display on appliances that are regulated. The label contains an estimate of energy usage and the approximate annual operating cost.

Efficient Refrigerator/Freezers

There are three arrangements for a refrigerator with a freezer: side by side, top freezer, and bottom freezer. The least energy efficient is the side-by-side arrangement. Protect the appliance from sunlight and do not place it near a heat source, such as the stove/range, oven, or dishwasher. The refrigerator appliance works to remove heat from inside. Any additional heat source adds to its workload. Clean the coils that dissipate this heat regularly to increase efficiency. Maintaining the freezing temperature is easier when thermal mass is high. So keep the freezer compartment as full as possible. It is hard to maintain temperature with just air inside. Adjust the temperature to the needed minimum. Freezer temperature should be 0°F to 5°F and the refrigerator should be 35°F to 38°F. When purchasing the appliance, avoid an automatic defrost model. Use of manual defrost may require some work, but the automatic defrost adds heat at regular intervals and increases energy usage.

Energy-Efficient Dishwasher

Look for the Energy Star label when purchasing a dishwasher. According to the DOE, dishwashers with the Energy Star label use at least 41% less energy than the federal minimum standard for energy consumption. Remember to operate your dishwasher only when full. Most of the energy used by a dishwasher is for heating water. For maximum energy efficiency, fill the dishwasher because it operates as if it has a full load. Many older dishwashers use about 8 gallons of water. Unless necessary, avoid using the heat-dry, rinse-hold, and prerinse features. Instead, use your dishwasher's air-dry option. Some new models have a feature called *Soil Sensing Technology*. This allows the dishwasher to sense the amount of debris and adjust the water used. Also avoid using a high-temperature setting or the heavy wash load setting unless necessary because these increase the electricity required for the cycle.

Along the line of lifestyle changes, hand washing dishes in the sink uses far less water. A typical modern dishwasher uses 6 gallons per wash. A sink would use between 1 and 3 gallons to perform the same wash. Additionally, there is virtually no electricity consumed with hand washing.

Cooking Appliances

The range, cooktop, microwave, and oven are all appliances that merit consideration when considering energy efficiency. Energy-efficient cooking saves in more ways than one. An example is the use of a microwave oven to cook. By more efficiently focusing the heat energy into the food, less heating load is added to the ambient conditions of the home, thus reducing the need for additional air conditioning.

Although electric cooking equipment is more efficient, the comparative cost of gas normally makes it more affordable. The question of gas or electric will present itself when you are considering which fuel to use. When comparing gas and electric, the aspect of cooking techniques cannot be ignored. Many professionals will only cook with gas. Personal preference should also be considered when deciding on which type of range to purchase.

Behavioral Changes

Consider the real need for all kinds of appliances. Appliances save valuable time and work and improve our lives, but not all are really needed. If possible, avoid the installation or use of unnecessary appliances. For instance, not installing a garbage disposal would be an obvious choice. Also consider the size of an appliance. The size of the appliance should match your needs.

Another way to save energy includes selecting cooking equipment to match the meal being cooked.

- A toaster oven may be all that is necessary for reheating leftovers.
- Microwave ovens cook in substantially less time and with less wattage.
- It might be more appropriate to use a small crock pot to cook a meal instead of a large pot over a burner.
- Select a pan material and shape that conducts heat effectively for the type of food being cooked.
- Use the appropriate burner size for the pot or pan being used. A pot that is too small for the burner yields wasted energy.
- Use lids to cover pots while cooking.
- Using glassware in the oven will allow the reduction of temperature as much as 25°F.
- Defrost frozen food in the refrigerator instead of the oven.
- Allow heated air to move throughout the oven cavity. Do not place foil or other material on racks.
- Cook double portions in order to cook less often.
- Preheat the oven only when ready.
- Use the oven light to check food instead of continually opening the oven door.
- Open the refrigerator door as few times as possible.

Individual energy use criteria are based on the average user nationwide and may not reflect your individual practices. A building science innovation created by Lucid Design Group includes a mechanism to monitor your energy usage in real time. Seeing your usage up front will allow you to more consciously adjust it. The Lucid Design Group illustrates its service:

> *In order to manage your resource consumption you must first measure it. But will you also be able to interpret it? Building Dashboard® helps you manage resource use effectively by displaying information in an intuitive way that non-technical users can understand. Building Dashboard® is further enhanced by the context of a growing community, one in which users can compare usage, share tips and strategies, and compete to improve the performance of their buildings and homes."*[3]

This technology represents resource savings through real time feedback. When we see how we are wasting energy, we can modify our behavior, reducing nonessential or optional energy usage. This is just one of the methods to making behavioral changes.

[3] *Lucid Design Group, inc., http://www.luciddesigngroup.com/*

BEFORE YOU DECIDE . . . REFLECTIONS AND CONSIDERATIONS

✔ Electrical energy brings us a higher standard of living.
- Lighting, appliances, equipment, labor-saving devices, and conveniences have improved our lifestyle but have increased the amount of energy we use.
- Bringing in natural lighting reduces the need for artificial lighting in the daylight hours and helps define a green home with energy savings.
- Skylights, light shelves, roof monitors, and similar methods illuminate a dark interior and make it into a usable space while avoiding the need to provide artificial lighting.

✔ Darkness at any time brings a need to provide artificial illumination.
- A green home makes use of the modern innovations that squeeze light out of a variety of sources that are more efficient than the normal incandescent bulb.
- Alternatives like fluorescent lamp bulbs, compact fluorescent lamps, light-emitting diodes, torchieres, cove lighting, and accent lighting provide an exceptional visual effect with limited expense.
- Demand responsive lighting, sensors, and dimmer switches also help reduce demand.
- Outdoor lighting that ordinarily provides security has been modified to illuminate only when motion sensors detect a presence.
- The misuse of lighting calls for behavioral changes of the occupant. Teaching the lost art of shutting the lights off when unused is a behavior change that makes the household truly green.

✔ The appropriate selection and use of appliances in the kitchen go a long way to having a green home.
- Refrigerator and cooking equipment account for a significant amount of energy use and misuse.
- The dishwasher and portable appliances add to the demand of electricity. Take advantage of the federal government's Energy Star program and energy label to guide a selection of an energy-efficient appliance.
- Behavior of the occupants can also make a difference. Properly used appliances represent an authentic green domicile.

FOR MORE INFORMATION

American Council on Energy Efficient Economy
http://www.aceee.org/

GreenBiz.com
http://www.greenbiz.com/

Integrated Building and Construction Solutions (IBACOS)
http://www.ibacos.com/

Lucid Design Group
http://www.luciddesigngroup.com/

Public Service of New Hampshire
http://www.psnh.com/

Chapter 9

WASTE MANAGEMENT

WASTE MANAGEMENT DURING CONSTRUCTION

As a society we produce and consume, but as a result of daily living we also waste. Waste results from inefficient use of material resources. In the construction process, waste means not using every part of the construction material specified for the job. In order to manage the amount of waste, a plan is needed to conserve all of the materials that arrive on the job site or are by-products of the construction process. Waste can be divided into different categories. One category is the physical state: solid, liquid, or gaseous. Another category is organic or inorganic. An important category to focus on is recyclability: whether or not that material is easy to reuse or recycle.

Consider the waste disposal methods for the items in the physical state category. For instance, solid waste is deposited in a landfill. Liquid waste could be collected for sanitary disposal elsewhere. Gaseous waste, such as exhaust or smoke, is not easily managed because it generally becomes part of the atmosphere. All waste might cause harm to the environment but in a different way. Solid waste can contaminate the soil and produce unsafe levels of methane and other pollutants. Liquid waste can infiltrate the ground and mix with groundwater, causing health hazards. Gaseous waste mixes with the breathable atmosphere and creates air pollution and respiratory problems.

When deciding on the waste disposal methods for organic or inorganic materials, it is important to keep in mind the organic material's reintroduction into the natural state. For instance, sawdust swept from inside a construction project would easily dissolve much like leaves and could serve as mulch. However, if the sawdust came from wood that had a preservative or fire retardant treatment, mixing it with your landscaping could introduce hazardous chemicals into your vegetable or flower garden. Similarly, you might be tempted to use cut pieces of lumber as firewood in your wood stove. Not knowing the chemical additives present in the firewood could lead you to accidentally introduce dangerous gases into your home.

Finally, whether or not the material is easily recyclable is the most useful classification for the purpose of building with the environment in mind. A green project is one that demonstrates minimum construction waste, such as the types illustrated in Figures 9-1, 9-2, and 9-3, and provides a method for recycling those items that otherwise would be headed to a landfill. You must do research in your area to determine the market for recyclables. If available, the obvious choice would be a recycling center that traditionally recycles paper, cardboard, glass, and plastics. A recycle building materials center could take your ordinary construction waste and resell it. For instance, if you purchased roofing shingles by the bundle (generally 3 bundles per square—100 square feet) and needed 632 square feet, you would have to buy 21 bundles because that is the only option. The leftover shingles seem inconsequential and represent waste. However, you might want to keep some for future repairs. But someone else could use those shingles for a shed or a porch. Virtually all construction waste can be reused in some fashion. Dimensional lumber can serve as blocking, brackets, or bracing even if they remain as shorter lengths. Masonry units such as block or brick can serve to enhance landscaping or garden fencing. Packaging such as paper, plastic, and glass could be recycled.

After considering all of these categories, if a solid substance is inorganic, inert, and not easily recycled, it is the most appropriate candidate for the landfill. Managing waste at a job site involves constant evaluation of the material and its effect on the environment.

Courtesy of iStock Photo

Figure 9-1 ■ Common building materials discarded as waste from a construction site commonly include wood framing and panel boards.

Courtesy of iStock Photo

Figure 9-2 ■ Sometimes construction waste includes short sections of framing lumber that could be used as blocking or backer material within a wall frame.

Courtesy of iStock Photo

Figure 9-3 ■ Construction waste commonly has residual pieces of the material used in the projects that are too small or the wrong size or shape to be conveniently used in the project.

Scope of the Problem

The problem of waste is clear upon the visitation of any construction job site. In virtually all construction projects, there are large roll-off dumpsters for anticipated storage or disposal of the leftovers from the project. Items that may be discarded include masonry, brick and block, metal, reinforcing steel, dimensional lumber of various sizes (some with nails), chunks of concrete, broken glass, broken pieces of drywall, roofing singles, empty paper and plastic containers that contained chemicals, tar, drywall compound, nails, packaging for materials, and other assorted products of the construction process.

In the United States alone, a staggering 136 million tons of debris from construction and demolition is created annually. With an approximate density of 50# per cubic feet, this represents over 200 million cubic yards of debris. Every year, construction debris adds a volume of 128 square miles 1½ feet deep. Imagine an area approximately the size of Philadelphia with 1½ feet of debris covering it each year.[1] Now, imagine if we reduced our waste by recycling what was immediately convenient: paper, aluminum, glass and plastic, and half of the construction debris we create. We could cut the volume of debris headed for the landfill significantly. We would also save on the cost and environmental impact of reproducing the items we would not otherwise recycle by using a few techniques and organizing the job site to make recycling easy.

Job Site Organization and Worksite Notices

An organized job site is a safe place to work, making it a responsible project. Most inspectors regard a neat and orderly job site as a quality building project and are more likely to pass inspections the first time. Being clean and efficient also allows the opportunity to recycle. But more importantly, it allows for the efficient and methodical cutting and assembly of components. Organization of a job site allows you to think before you act. If the lumber is organized by dimension and approximate length, your search will be quicker and you will not inadvertently use a longer piece and thus create waste. The job site organization will be unique and influenced by site considerations like size, shape, terrain, ingress, and egress.

Elements of Good Job Site Organization

There are numerous things to consider when you plan a construction project. Financing, material selection, subcontractors, timing, project time lines, cost, and weather are just a few of the aspects. Most contractors will not evaluate and establish job site organization until they arrive with a crew and begin unloading tools, power cords, and generators and work with the conditions presented to them by the location of materials and the existing terrain. Many good carpenters know instinctively the best layout for the conditions. They develop this instinct based on years of experience.

With so much to be gained by simple planning, it makes sense to start with a plan that results from careful consideration of the many factors that affect efficiency and your goal of a truly green home. Here are a few elements that should be considered when making a plan for job site layout.

[1] *Environmental Protection Agency, http://www.epa.gov/*

1. How does topography affect your layout?—Mostly flat terrain presents few problems, but sloping land can be difficult if materials must be hauled uphill. In addition, materials stored in lower areas might collect rain runoff and damage lumber. Trees may help if you need a *screen* to conceal the work because of zoning or subdivision rules.
2. How does utility service affect your layout?—During construction, you will need electricity and water. Normally, the utility company will provide temporary utilities for the construction project. For electricity, it is normally up to you (or your electrician) to provide a temporary electric service that can be connected. When the home is complete, a temporary electrical service will normally be where the permanent electrical meter and service disconnect will be located. Consider if the temporary electric service is convenient or causes problems during construction.

   The same is true for temporary water service. It needs to be convenient for construction work, yet far enough away to avoid hazards or accidental damage to the meter or water valves. Find out if there is a choice as to where the water service meter will be located.
3. Where will your construction site workstation be situated?—This workstation can include tools like the table saw, miter saw, compressor, generator, nail gun, and the associated nails, staples, blades, hand tools, and cleanup materials. Adequate room is needed for the necessary tools. Weather should also be a consideration included when deciding where to set up the workstation. Care needs to be taken to be sure tools will stay out of wet conditions. Location is another factor for the workstation. Finding a convenient spot that allows for rapid setup and use of tools is important.

Building material storage area

Where you store your materials depends on several factors. You want the materials to be close to your work area in order to be efficient. They must be in a place that is clean and dry to avoid damage and deterioration like that shown in Figure 9-4. They may need to be out of sight because of zoning restrictions or subdivision regulations. Hiding them from public view may help reduce pilfering as well. Elevate large banded stacks of dimensional lumber on pallets to avoid water runoff damage. In any case, be on site during delivery to ensure that your materials are carefully unloaded and placed exactly where you want. Try to avoid the wasteful result of a delivery to an improper location.

Local ordinances

Construction of your house will be disturbing to your neighbors. There is dust, debris, noise, vehicles, and other associated pollution that goes along with the construction process. Your neighbors could be protected from this unwelcome intrusion. There may be rules and regulations that affect your work. For instance, your work may be limited by the clock based on a zoning or subdivision rule. The available time for work may be limited to 8:00 A.M. to 5:00 P.M. There may be a noise limitation based on decibels. You may be required to build a screen to shield your work area from the public. You may need a special permit to engage in construction even though you have a building permit for the house. Be sure to ask about these stipulations when you apply for a building permit.

Postconstruction restoration

Site restoration will take place once the construction is complete. In some cases, the work station could transition to the interior of the house after the roof is installed. Keep the future landscaping in mind when deciding on the location of the workstation. Long-term damage to

Courtesy of iStock Photo

Figure 9-4 ■ A poorly thought-out location for storage of building materials could result in premature deterioration or saturation with water. This could be the result of a lack of communication between the delivery driver and the builder. Be on site during material delivery to avoid this type of problem.

the soil can be avoided by establishing basic environmental soil erosion control measures such as the following:

- A siltation fence around the perimeter prevents soil erosion from migrating beyond your job site.
- Avoid mud or dirt tracked onto the street or road by establishing mitigation with gravel bed or a similar measure. This mud-tracking pad retains the dust that would otherwise be deposited on the road or street.
- Spray water or a similar measure to control dust and airborne erosion.
- Protect storm water drainage inlets. This is a measure that protects storm water or runoff. A solution could be as simple as installing filled sandbags over inlets to filter the runoff.
    - A storm drain basin can be established with rip rap or aggregate held together with one of many ways, including fencing or geotextile. This rip rap prevents soil erosion caused by rapid runoff of water.
- A restoration plan should be in place to restore the construction site to the original environmental condition. This may include plantings and grass or wind wildflower seeding.

Framing Plan

A framing plan includes the details of assembly for the components of a building. In the case of a wood frame building, the plan and details would note the assembly and placement of the wood frame, floor joists, roof rafters, blocking, studs, plates, headers, cripple, trimmers, and so on. The plan, like the one shown in Figure 9-5, would note the size, center spacing of each repetitive piece, and connection details to supporting elements.

Most projects built of wood framing by a carpenter have some type of framing plan. If not a formal plan, at the very least the carpenter will organize the plan mentally. If this is your first time building a wall frame it is most helpful to have a drawn-out plan to which you can refer as each step is taken.

Cut List

The value of having a framing plan is the proper management of the lumber and the cuts needed to create the necessary sizes. A cut list is the complete list of various pieces of lumber that must be cut to form an assembly: a wall, roof, or floor frame.

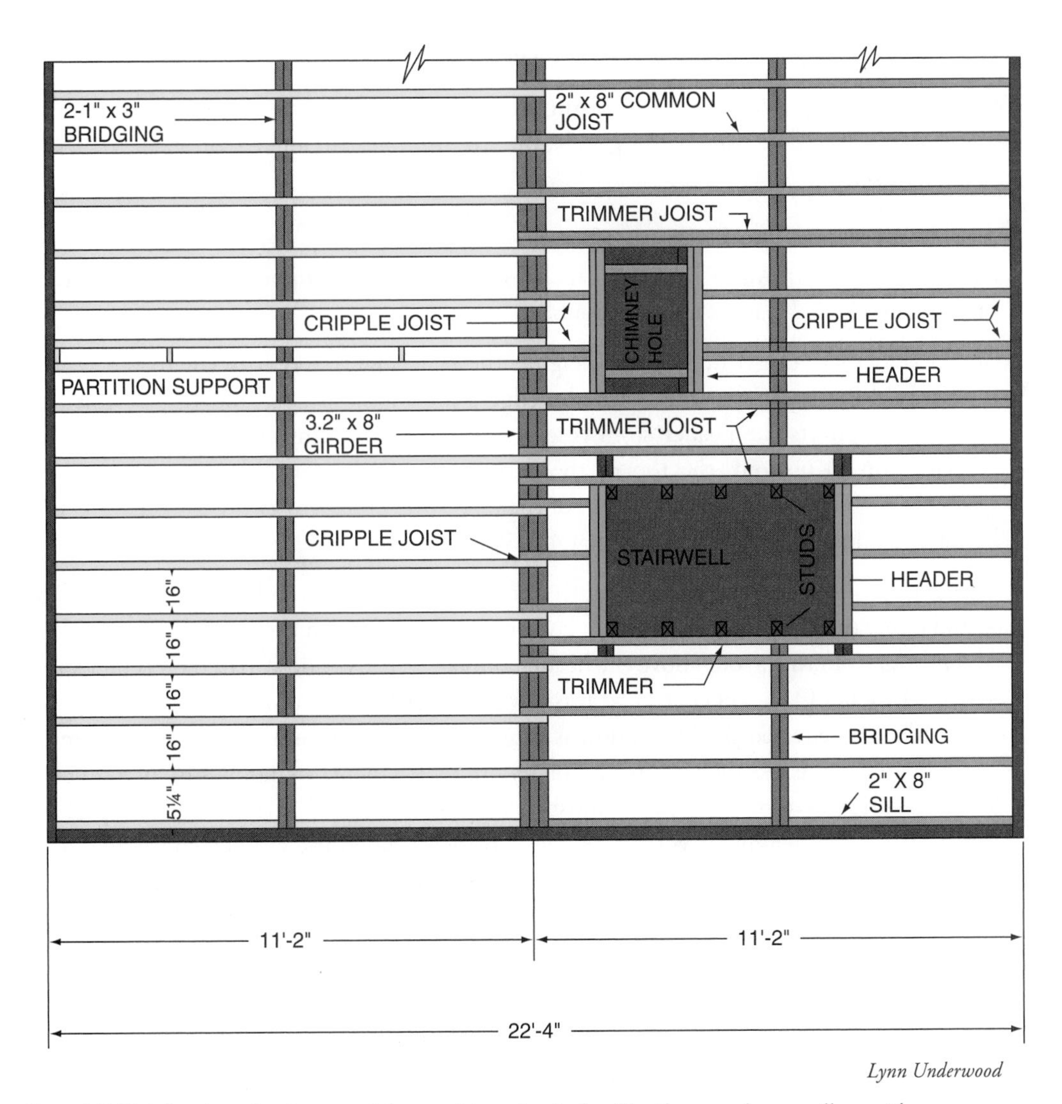

Lynn Underwood

Figure 9-5 ■ A framing plan is essential to avoid waste. A plan like the one shown will provide you or the carpenter with valuable information on the exact sizes of repetitively cut joists and rafters.

For example, a simple wall frame that is 10 feet long and 8 feet high could include one centered window in the wall. The window is 4°4°. The wall will be made from 2 × 4-inch lumber. The studs are spaced 24 inches apart. To calculate the lengths needed for each piece of the wall frame assembly, follow these steps:

- *Step 1*: Calculate the wall plate: 3 – 10 feet long.
- *Step 2a*: Calculate the number of studs: one on each end and one every 24 inches.

  10/2 = 5. 5 + 1 + 1 = 7
- *Step 2b*: Adjust the stud count to reflect the window. The window is centered in the wall frame and is 4 feet in length and height, leaving 3 feet of wall length on either end. There are now two wall sections within this 10-foot length, each 3 feet long and including a connecting header. With a wall section on each end and one in the middle, the total number of 92⅝-inch standard length studs is 6 (one on each end and one at 24 inches away from the starting stud).
- *Step 3*: Calculate the stud length. With one bottom plate and two top plates, a stud within a frame for an 8-foot wall height will be

  (8') (12"/Foot) = 96" − 3(1½") = 91½ inches.

To accommodate an 8-foot ceiling and assuming about an inch to accommodate a ceiling drywall and finish flooring, a standard *stud* precut length of 92⅝ inches is traditionally used.

- *Step 4*: Calculate the header length. If required for bearing to support the window opening, a header needs to be 48 inches + 1½ inches on each end (for support by a trimmer stud) or 51 inches long. Headers must be doubled.
- *Step 5*: Calculate the trimmer stud height. The full height stud adjacent to the header is called a king stud. It connects to the edge grain of the header and the trimmer, or jack stud; supports the header from below; and ties to the king stud. If the window is centered top to bottom, the top of the window or the bottom of the header will be [(8' – 4')/2] +4' or 6' from the bottom of the floor. Therefore, the two trimmer studs will be 6 feet minus 1½ inches (bottom plate) or 70½ inches long.
- *Step 6*: Calculate the number and length of cripple studs. The cripple studs are those that would have continued the normal stud layout that changed with the window. They merely fill in the gaps in the layout above the header and below the window. Based on center spacing of 24 inches O.C. (on center), at least two cripples below the window and two above the header are required. If the window is centered top to bottom, the lower cripples must be 24 inches minus 1½ inches (bottom plate) minus 1½ inches (window sill), or 21 inches long. The cripples above the header must be 24 inches minus 3 inches (double top plate) minus the header height. If you chose a 2 × 6-inch header, the cripple length would be 24 inches minus 3 inches minus 5½ inches (nominal size of 2 × 6-inch header), or 15½ inches long.

The cut list would then look something like this:

Plate: 3 – 10 foot long (Note: includes two top plates and one bottom plate)

Studs: 6 – 92 ⅝ inches long (normally precut to length)

Header: 2 – 2 × 6 inches, 51 inches long (only required for bearing condition)

Trimmer Stud: 2 – 70 ½ inches

Cripples: 2 – 21 inches

2 – 15½ inches

Central Cut Area

During your layout of the work area, you will decide where the various pieces of lumber will be cut as outlined in the previous section. The organization of the site will set out how efficient and accurate your work will be. Being near the material storage and adjacent to the home site is essential for efficiency. Position your cut area in proper relation to the material storage area.

Drywall

According to the National Association of Home Builders, approximately 15 million tons of drywall is produced every year in the United tates. About 12% of that becomes waste and is stored in landfills. Most drywall waste comes from the new construction process (64%).[2] Drywall products are generally produced in 4-foot-wide sheets that are 8 to 16 feet long. The thickness varies from ¼ inch to 1 inch with ½ inch being the most common. With a 9-foot ceiling height becoming more popular, some drywall sheets are 4½ feet wide (installed horizontally). Drywall waste can be reduced by:

- Using lengths of walls and height of ceilings consistent with drywall dimensions
- Selecting custom-sized sheets for nonstandard walls and avoiding standard size sheets that would necessitate breaking pieces
- Carefully measuring dimensions before cutting
- Placing drywall scraps within the wall cavities of interior wall frames to avoid landfill waste

Recycling Plan

At any job site, there is an enormous volume of fully recyclable waste that is considered to be trash only because it is inconvenient to recycle. If only a third of all construction waste was recyclable, over 40 million tons of waste would be removed from the landfill every year. When recycled that same product would eliminate the need to manufacture a new product, thus doubling the savings. The recycling plan does not have to fit within the same municipal or regional plan. You may find that a recycling center requires you to bring the materials to them.

Many types of recycling materials are routinely found at an average home construction site. These include cardboard, wood, metal, plastic, drywall, and shingles. The average volume of debris for each is staggering when you consider the number of houses built every year.

- 600 pounds of cardboard are discarded from the average homebuilding site. Those are 20 cubic yards of debris that are deposited in a landfill.
- 3,000 pounds of wood are disposed of while building the conventional wood frame home. Those are 11 cubic yards of wood that will fill a landfill.

[2]*National Association of Homebuilders,* http://www.nahb.org/

Courtesy of iStock Photo

Figure 9-6 ■ Drywall is a major source of construction waste. Much of the waste is unnecessary. Drywall is now able to be recycled.

- 150 pounds and 1 cubic yard of metal will become trash although it is easily recyclable.
- 150 pounds and 1 cubic yard of plastic gets hauled away to the dump. Many plastics are recyclable and could save valuable space in a landfill. As much as a ton of drywall is discarded in the average homebuilding project. That represents 6 cubic yards of waste. Drywall, such as that illustrated in Figure 9-6, is the principal wall material used in the United States for interior purposes.
- Asphalt shingles are another serious contender for construction waste. Approximately 11 million tons of waste asphalt roofing shingles are generated in the United States per year.

BEFORE YOU DECIDE . . . REFLECTIONS AND CONSIDERATIONS

✔ Waste not . . . Want not. The old adage reflects the mission of recycling proponents.

- The waste we produce still has to be stored by us in a waste disposal site.
- A construction project is filled with waste of all types.
- Avoiding waste and using as much of the building material as possible are marks of a *green home.*

- ✔ There are many ways to decrease or even eliminate waste.
  - Careful house planning in the design stage is important.
  - Manage the construction project by building an efficient production process with notices to workers regarding waste and recycling. Add organizational elements like:
    - A detailed framing plan for use of all available lumber
    - A detailed cutting analysis to avoid accidentally wasting lumber
    - A central cut area to consolidate operations and avoid duplicate waste
    - Using care when making cuts on all material, including wood framing, sheathing, and drywall
    - Arranging the workstation to reflect topography
    - Reducing damage to material by nature with proper storage
  - Add a plan to salvage all materials that can be reasonably recycled.
- ✔ Restore the construction site when complete instead of leaving a scar on the landscape.
- ✔ Proper management of the project is essential to pull all these components together.

FOR MORE INFORMATION

Environmental Protection Agency (EPA)
http://www.epa.gov/osw/inforesources/news/speeches/
http://www.epa.gov/osw/hazard/generation/

National Association of Homebuilders: Builder's Field Guide
http://www.nahbrc.org/bookstore/

Chapter 10

WATER EFFICIENCY

WATER EFFICIENCY

Between 1950 and 2000, the United States population nearly doubled. During the same period, demand for water nearly tripled. Each man, woman, and child uses an average of 100 gallons of water every day. With the ease of turning on a tap, water is accessible. Most of us give precious little thought to the energy and resources expended to make that happen. Most people know that hot water creates a demand for energy, but delivering cold water to your home does as well. The American public water supply consumes 56 billion kilowatt-hours per year to deliver fresh water to the faucet. That is enough electricity to power more than 5 million homes every year. Water conservation saves on environmental impact in many ways. If 1% of American homes replaced an older toilet with a low flow valve, the savings would be 38 billion kilowatt-hours of electricity. That is enough to supply the electrical demand of over 43,000 homes for a month.[1]

Demand for water creates the need to build dams, dig wells, and deplete water from natural reservoirs such as lakes and rivers. The demand for water creates conditions that draw water across ground sources, collecting pollutants and depositing them in our drinking water. Excessive pumping of groundwater from these sources can cause changes in waterways such as streams, allow infiltration of saltwater into freshwater aquifers, increase polluted runoff into natural water supplies, and create the need for more dams that cause other environmental issues.

For some, it is easy to rationalize that the hydrologic cycle will supply clean, safe drinking water that will always be there when we want. A recent government survey indicated that 36 states anticipate potable water supply shortages within their jurisdiction by the year 2013. Increasingly, the necessity for conserving water is becoming critical. Green homes use water very efficiently. There are numerous ways to conserve water in your new green home both inside and out.

[1]*Ipswich River Watershed Association,* http://www.ipswichriver.org/

The benefits to water efficiency are many. Saving water has a compound effect. Not only are you saving water, some sewer purveyors charge based on the rate of water consumption. No matter where the water is going, you are paying for it to go into the sewer system. Conserving water will create less demand for additional sewage treatment facilities. Also, there is an energy cost to bringing water to your home. Saving water reduces that energy demand. Almost half of your water is heated. Wasting hot water costs the energy expended to heat that water. Reduced demand for water will preserve existing water sources that also support biodiversity in the environment.

WaterSense Label

WaterSense is a partnership program sponsored by the Environmental Protection Agency (EPA). The EPA established WaterSense as a voluntary standard for flow rates from a variety of fixtures, including kitchen and bathroom faucets, showers, and others. This is a program that evaluates and identifies certain fixtures as being water misers. With this label, consumers can evaluate the most efficient fixture for water use. The label appears on toilets that use 20% less water than standard models. A bathroom faucet will use 30% less than that currently mandated by law (2.2 gpm at 60 psi)—a maximum of 1.5 gpm.[2]

Indoor Water Use

We drink water, bathe in it, cook with it, cool our drinks with it (ice), wash our clothes, sanitize living surfaces (such as mopping), and care for plants. We pay for it to be delivered and then pay again when it is taken away down the drain (sewage bill). Although reasonable use is to be expected, wasted water is definitely not part of a green home. Indoor and outdoor water conservation for a green home has many avenues.

Pressure Reducing Valve

High water pressures can waste a lot of water. The International Residential Code (IRC) only requires a minimum working pressure of 40 pounds per square inch (psi). Reducing overall water pressure to your home is a water conserving measure that can save water and cut your water bill. The water company will advise you of the water pressure at your home. If it is above 70 psi, consider installing a pressure reducing valve. The International Plumbing Code and the IRC require that a pressure reducing valve be installed if your water pressure is over 80 psi. This device reduces the pressure applied to deliver a more manageable level that reduces the volume of water to your home. The reduction in pressure will reduce the volume of water discharged by faucets and reduce overall water usage. An expansion tank is required when a pressure reducing valve is installed.

Water Leaks

Water leaks should be repaired as soon as possible. The drips may not seem like a serious loss of water, but a single drip every second amounts to 3,000 gallons per year. In a bathroom or kitchen that leak could be hot water as well. Check for leaks in the water supply piping system routinely. Look for signs of wear and tear around faucet valves. Replace or repair faucets, pipe,

[2] *International Residential Code (2006), International Code Council, Inc., Table P2903.3, and the Environmental Protection Agency, http://epa.gov/watersense/*

or other water supply systems as needed. One sure way to check for leaks is to turn all water valves off, read your meter, and check again in a few hours to see if the meter registers any change. A leaky toilet can waste about 200 gallons of water every day. The mechanism for ensuring that polluted sources such as a toilet bowl *do* not contaminate the potable water in the tank may need to be replaced. A way to verify if there is a leak is to put some food coloring in the tank water and see if any seeps into the bowl (without flushing of course). This may take several minutes, but do not wait too long because this coloring agent may stain the inside of the tank (and bowl).

Clothes Washer

We wash our clothes for sanitary reasons and to present a civilized appearance. That clean appearance comes at a cost. Almost one-fourth of all domestic water use is for washing clothes. Washing clothes uses about the same water as all toilets in the home. Making a change to a high-efficiency washing machine can conserve a large amount of water. Traditional machines use between 25 and 50 gallons of water per load (wash and rinse cycles). Front-loading-type or top-loading nonagitator-type machines use about half that water amount. There are other models that are more conservative. A way to make each wash efficient is to ensure that wash loads are full.

The Consortium for Energy Efficiency (CEE) is a nonprofit corporation that promotes initiatives and energy-efficient products and services. It establishes the efficiency levels for energy usage of appliances and equipment within a home. Through the Super-Efficient Home Appliances Initiative, the CEE promotes the manufacture and sales of energy-efficient clothes washers.[3]

Bathrooms

Bathrooms consume around 25% of all domestic water used. There are at least three water supply sources in most bathrooms: tub or shower, toilet, and sink. Each fixture comes with its own ways of conserving water.

Bathtubs

There is nothing more comforting than a hot soak in a bathtub. There is nothing in the home that uses more water at one time either. A full bath takes around 70 gallons of water, whereas a 5-minute shower uses between 10 and 20 gallons. Therefore, choose a brief shower over a bath most of the time. For those times when you must have a bath, install the stopper before you start the water and then adjust the bathing temperature to suit your needs using the hot and cold faucets.

High-efficiency shower heads

The average shower is a major source of domestic water use in the United States. Approximately 17% of residential indoor water use, over one trillion gallons of water, is consumed in the shower each year.[4] Many people believe that a decreased flow rate will still satisfy the need for a good shower. Such an acceptable flow rate is yet to be specifically defined. The EPA is partnering

[3]*Consortium for Energy Efficiency, http://www.cee1.org/*
[4]*EPA WaterSense®, http://epa.gov/watersense/*

with several standard developing agencies to identify a rate that is significantly lower than the prevailing shower heads.

Two basic types of low-flow showerheads are the aerating flow and the laminar flow. Aerating flow showerheads mix air with water, forming a fine spray. Laminar flow showerheads create small streams of water. In a humid climate, a laminar flow showerhead is better because a stream of water will not make as much steam. Twenty years ago, many showerheads had flow rates of 5.5 gpm or more. Do not use a recycled shower head if the age is in question. Most modern shower heads have flow rates of less than 2.5 gpm.

Shower drain heat recovery device

Hot water goes down the drain just like cold water. The thermal energy used to heat the water for bathing or cooking could still be used. Drain water heat recovery devices save this otherwise lost heat energy. If recovered, this energy can be used to preheat the cold water headed to the water heater. Although the water may go down the drain, these systems reduce the cost for heating water. The method for reclaiming the heat is through a heat exchanger. Heat exchanger configurations vary in terms of pipe layout, orientation, and design. They also vary in cost and the amount of energy savings achieved. Several manufacturers have developed approaches to the device. They all seek to reclaim energy for an alternative use. The reclaimed energy comes from appliances or fixtures that hold or transfer heated water in some manner. A common source is the shower or bathtub. A dishwasher has similar heated water that would be ordinarily discharged through the drainage system.

High-efficiency bathroom faucets

Bathroom faucets are responsible for around 15% of domestic water use. Earlier faucets delivered between 4 and 7 gpm. Because of federal law, bathroom lavatories now have a flow rate of not more that 2.2 gpm. Many available ones discharge less than 2 gpm. Specialized faucets that earn the EPA's WaterSense label restrict flow rate to less than 1.5 gpm. A faucet aerator can reduce the flow rate even further.

High-efficiency toilet

Toilets are a top user of domestic water, consuming over one fourth of all water in a home. Older toilets use several gallons for a single flush. Steady improvements in technology have lowered the consumption rate for this fixture. Under federal law, new toilets cannot exceed a demand of 1.6 gallons per flush (gpf). High-efficiency toilets have a maximum threshold use of less than 1.3 gpf. The WaterSense label is used on these toilets that are certified by independent laboratory testing to meet rigorous criteria for both performance and efficiency. The WaterSense high-efficiency toilet must pass these performance tests so that it performs well and saves water.

The amount you can save with simple practices and efficient fixtures is astonishing. Over the course of your lifetime, you will flush a toilet about 140,000 times. With a high-efficiency toilet, you can save 4,000 gallons per year and 300,000 gallons in a lifetime.

Kitchen

Kitchen faucets are a source of waste in potable water. Cooking and cleaning, by necessity, involve water. Depending on the water pressure, older faucets supplied up to 7 gpm. We have gotten a lot better over the years. High-efficiency kitchen faucets are now available at a flow rate of less than 2.75 gpm. There are some that limit flow to 2.5 gpm. Adding an aerator to a kitchen faucet reduces the rate even more. Aerators blend water and air and reduce flow rate but do not affect the prevailing water pressure.

Consumer Behavior

Consumer practices are among the greatest single source of water savings. Even with all the water-saving technology that we have at our disposal, the human factor is still a significant reason for misuse of this life-giving resource. Simple changes in our behavior can profoundly affect our use of water. The average home spends $500 annually on water. Based on this average, making changes in technology or behavior could save as much as $170 each year. Some of the following tips are behavior changes that may challenge you at first but will net water savings if followed strictly.

- Water plants wisely. For indoor plants, use a watering spout and do not overwater.
- Use full loads to wash clothes and dishes.
- Turn the water tap off when brushing your teeth or shaving.
- Look for the label WaterSense Meets EPA Criteria that exceeds regulatory standards by 20% to 40%. Each manufacturer's product has to pass certain criteria to earn this label. A third-party certification is necessary to earn the label.

Outdoor Use

Even in a green home, the living yard, garden, flowers, and other landscaping still need to be irrigated. We wash our cars, fill our swimming pool or hot tub, wash off our driveway, clean our siding, and wash our windows among other things. We pay for water to be delivered and give very little thought to the alternative approaches to traditional outdoor uses we make of water. Outdoor water conservation for a green home is essential.

Landscaping

Landscape should include decorative rock or stone, native plants or those that that are drought resistant. Xeriscape is a type of landscaping, like those illustrated in Figures 10-1 a and b, that does not require supplemental irrigation in order to survive and flourish. Originally intended for arid climates lacking precipitation, this type of landscaping is used more commonly in green homes. Many varieties of plants tolerate arid conditions and are drought resistant. These plants are not just desert-type landscaping such as cactus or thorny bushes. However, a low-tech solution to landscaping that avoids the need for irrigation is to select plants that are native to the area. If they live without active irrigation in nature, you should not have to add too much supplemental water to keep them alive.

Rocks

Rock is appropriate for landscaping in arid locations. Cleaned and sorted by size, rock makes an excellent landscape choice in areas where few plants grow naturally or where conditions are such that irrigation is essential to maintain a living landscape. There are numerous variations in using stone and rock in landscaping. Many include the use of plants. There is common stone, river rock, flagstone, moss rock, and even boulders.

(a)

(b)

Courtesy of iStock Photo

Figure 10-1 ■ A large part of household water use is for landscaping. One alternative to conserving water is to develop landscape that relies on a minimal amount of water. Xeriscape plants rely on normal rainfall, not on supplemental irrigation.

Courtesy of iStock Photo

Figure 10-2 ■ Flagstone installation is a beautiful landscaping feature that avoids the need to irrigate with water.

An example of the effective use of rock as landscaping is the flagstone that serves as a walkway in Figure 10-2.

There are some considerations for landscaping with stone and rock. First, try to use stone or rock that is native to your area so it does not look out of place. Second, when selecting a large size rock, pick one with irregular detail that attracts the eye. Third, select a color that enhances your home and does not compete with it for attention. The landscaping should accent the home and not be an exclamation point. For a green home, it is important to avoid creating a hardscape that prevents rain infiltration, so leave drainage potential.

Timer for irrigation

No matter what you are irrigating, it only needs so much water to survive and flourish. Using more water than necessary is wasteful. A landscape irrigation system feeds water to those plants that need it and avoids overwatering if a timer is added. A timer or programmable system will apply the amount of water needed at the optimum time.

Drip irrigation

Drip or trickle irrigation is the introduction of small or modest amounts of water at a very slow rate with low pressure over a long period. For the concept to work optimally, the delivery rate should be the same at all areas without a wide variation. A soaker hose is an example of this type of irrigation and is effective because it can be targeted directly on the root area of the plant to avoid some loss realized by other irrigation methods from evaporation.

Gray water usage

Gray water is domestic wastewater that has been used in relatively clean processes within your home such as a kitchen sink, bathroom shower or bathtub, lavatory, or even the dishwasher or clothes washer. The wastewater normally would go down the drain and into the sewer system and mix with black water. Using a system like the one depicted in Figure 10-3, gray water would be diverted to a tank and either fed to landscaping or retained until needed. Water of this type is better when used quickly and not stored for a long period.

Consumer Behavior Practices or Technology Advances

Consumer practices are among the greatest single source of water savings. The human factor is still a significant reason for misuse of water used outdoors. Simple changes in our behavior can profoundly affect our use of this natural resource. The average home spends $500 annually on water. Based on this average, changes in technology or behavior could save as much as $170 each year. Some of the following tips are behavior changes that may challenge you at first, but will net water savings if followed strictly.

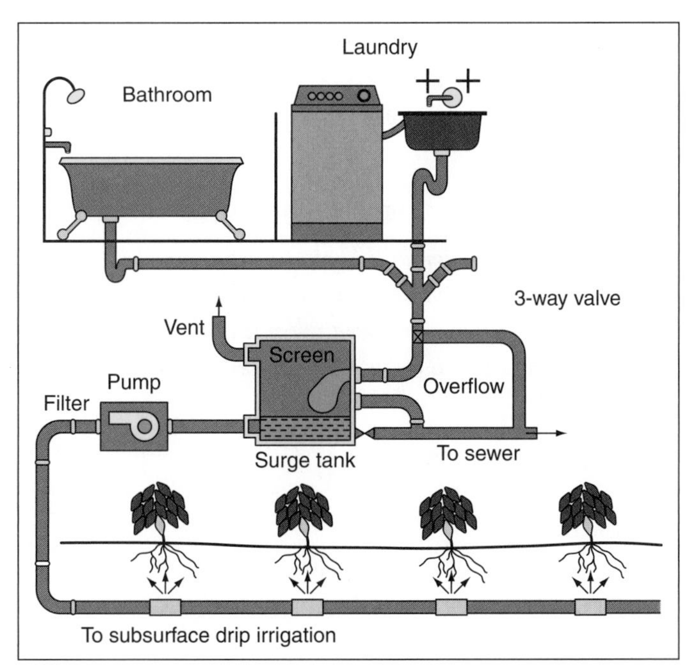

Lynn Underwood

Figure 10-3 ■ Water used in the home, except water from toilets, is called gray water. Dish, shower, sink, and laundry water comprise over half of all household water used. This water, otherwise sent down the drain, may be reused for other purposes, especially landscape irrigation.

- Water plants early in the morning. This reduces evaporation losses that occur in the heat of the day.
- Buy sprinklers that produce droplets and not mist. Use a soaker hose or trickle irrigation for trees and shrubs.
- Use sprinklers to water lawn and garden only. Make sure that you are not watering the sidewalk, driveway, or street.
- Do not overwater your landscaping. When we irrigate, we tend to imagine that one size fits all for plants. Different plants have different watering needs. Read up on these and water appropriately.
- Do not overfertilize. This will require more water to keep the grass alive and will add unnecessary pollution to the groundwater.
- Depending on the type of grass lawn you have, raising your lawnmower blade at least 3 inches will allow the grass to grow. In common varieties of grass, this will allow a deeper root system where the soil holds more moisture. Less watering will be needed.
- Control weeds that otherwise would steal the water from your landscaping.
- Plant vegetation that is native to your area. These native plants tend to need less water because they are accustomed to the natural rainfall.
- Install mulch around trees and plants. This retards evaporation and protects the base of the plant as well.

Permaculture and Water Harvesting Principle

The Permaculture Institute in New Mexico provides course work and training in certain self-sustaining techniques including water harvesting. Following is its purpose:

> *Permaculture is an ecological design system for sustainability in all aspects of human endeavor. It teaches us how to build natural homes, grow our own food, restore diminished landscapes and ecosystems, catch rainwater, build communities and much more.*[5]

An explanation of the principles or rainwater harvesting is also given:

> *In permaculture we strive to design buildings and landscapes to absorb rainwater. This is not only a good idea for dry climates, but is also very important in places with plentiful moisture. Why? Rainwater is best used when it is allowed to infiltrate the soil. There it is available to plants, it is cleansed and enters the groundwater or returns to the hydrological cycle.*[6]

Rainwater harvesting is also an alternative to designing conditions that must control runoff water to avoid erosion. There are numerous options for using rainwater in landscaping and other areas around the home.

[5] *Permaculture Institute, http://www.permaculture.org/*
[6] *Ibid.*

BEFORE YOU DECIDE . . . REFLECTIONS AND CONSIDERATIONS

- ✔ Next to oxygen, water is the most essential ingredient to life on our planet.
  - Although the planet is 70% water, most of it is not potable.
  - Only about 5% of all the water on Earth is clean enough to drink.
- ✔ Conservation of this vital natural resource is a hallmark of a green home.
  - Both indoor and outdoor water use demand can be reduced by technology or practice.
    - Use water-efficient equipment and fixtures.
    - Efficient design offers extremely low usage fixtures such as toilets, faucets, shower heads, clothes washer, and bathtubs.
    - Most of these extremely low-use fixtures are identified by ratings labels established by the EPA.
      - The WaterSense label marks a product that has significantly reduced water flow.
- ✔ Another method is to reuse water by adding a gray water system to your house to allow for a second use of water that has been used for light washing, rinsing, or cooking.
- ✔ Outdoor water use can be reduced by selecting native or drought-resistant plants. This method of landscaping is called xeriscaping.
- ✔ Another method is to spot irrigate in a way that avoids water loss by evaporation.
- ✔ Water harvesting is an excellent way to protect water resources.

FOR MORE INFORMATION

Brad Lancaster: Harvesting Rainwater
http://www.harvestingrainwater.com/

Consortium for Energy Efficiency (CEE)
http://www.cee1.org/

Environmental Protection Agency (EPA) WaterSense
http://epa.gov/watersense/docs/

Ipswich River Watershed Association
http://www.ipswichriver.org/

Permaculture Institute in New Mexico
http://www.permaculture.org/

Chapter 11

RENEWABLE ENERGY

ALTERNATIVE ENERGY SOURCES

Energy is commonly defined as the ability to do work. Defining energy is difficult because it exists as two states—potential and kinetic—and to some degree, it resides in virtually everything. Another way to reference energy is as a mathematical abstraction: a measure of its effect on matter. Energy can be thought of as a measure of work or, simply, the measure of change. When energy is introduced into a system, change is the result. There are many different manifestations of energy, including mechanical, thermal heat, light, chemical, electrical, solar, and nuclear. Heat energy can move in several ways, such as conduction, convection, and radiation. However, although physicists argue that energy is an indefinable phenomenon, for the purposes of using energy responsibly, it can be regarded as the physical force that makes things happen.

Energy is essential for our homes and everything that touches our lives. We need energy to heat and cool our homes and offices. We use energy for lighting as well as for the comforts of home that we enjoy, such as entertainment and communication, and for business purposes that fill our lives. In our current society, energy comes to us in one of two conventional sources: fossil fuel (in the form of petroleum products such as natural gas, LP gas, or home heating oil) and electricity (that is produced from a variety of sources, including fossil fuel). Both of these place a significant demand on the environment in many ways, causing either direct or indirect effects. The direct effects on the environment include the by-products of production, the impact of delivery, and the result of use. The indirect effects are similar but are spin-offs of the direct effects. These are things like climate change, changes in water flow, increased erosion, and pollution. Although we have covered the effects of traditional energy consumption, there are alternative energy sources that leave a smaller environmental footprint or, in some cases, almost none at all.

Alternative energy is not new. In fact, before the discovery of traditional sources of energy, some of these alternative sources were considered mainstream. For instance, even as recently as the 1950s before clothes dryers were common, most clothes were washed at home yet dried outside on a clothesline. The quality of the service provided by the sun is sought after by modern products. Advertisements for chemical additives in modern dryers often refer to the fresh smell and crisp feel of being just like sun dried. Wind energy was harnessed several thousand years ago. The earliest ships used the wind to travel. Wind drove mills that ground seed products into wheat meal. Before the 1800s neither electricity nor oil had been discovered and pioneers made do with the prevailing energy sources in a variety of ways: sundials for timekeeping, candles for light, and coal or wood for heat. Domesticated animals provided transportation as well as brute strength for agriculture, manufacturing, and food production. Hydropower from rivers first powered mechanical sawmills and grain mills.

In the last century, our civilization discovered or created processes that are convenient sources of energy, including electricity. Combined with inventions in transportation and other amenities, our society has come to rely on these expedient sources of energy. We now rely on the convenience of flipping a switch for light or entertainment. We think nothing of flying across the country for leisure. We drive golf carts for a sport that would otherwise provide exercise. Without considering the consequences, we drive an SUV all around a parking lot to find the closest space to a store when parking farther away would add exercise to our regimen and avoid gasoline waste. We ignore the impact we have on the environment by so many frivolous gestures that consume energy and, in sum, have an adverse affect on the Earth.

In recent times, the cost of energy has escalated enough to get our attention and caused us to move toward alternative sources of energy production. But motivation is needed to encourage the use of an alternative energy source. We need to move away from dependence on nonrenewable petroleum products or other energy sources that bring harm to the environment. Alternative energy systems use renewable natural resources to generate and use electricity to heat or cool the home. This chapter discusses a few of the kinds of alternative sources that you may consider in your green home.

However, renewable energy systems tend to be expensive. It is a smart principle to always reduce your energy demand before you endeavor to generate your own. If you reduce your overall load through all the measures mentioned in the previous 10 chapters, you can meet those demands with the smallest possible system. Energy waste is unnecessary and many times caused by energy losses in the system as well as excessive consumption. Proper energy management, energy efficiency, and minimization of loads should be the first steps toward energy generation. Eliminating practices like using surge protectors that remain on all the time, leaving lights on when not needed, playing the radio or TV loudly, and similar careless actions are behavior changes that your family can make.

Active Energy Production Systems

Active systems generally refer to those systems that are mechanical in nature or use an active process to facilitate the use of an alternative energy production system. A passive system is one that facilitates energy production (or transfer) without the use of mechanical processes. For instance, photovoltaic cells that generate electricity could be regarded as an active system because it relies on a mechanical system to transfer solar energy to electricity. A thermal mass wall adjacent to a south-facing window that absorbs solar energy and that has openings at the top and bottom to create a convective loop would be regarded as a passive heating system. Other chapters have discussed passive systems, including solar heating and water heating design. This chapter limits the focus to energy production or energy use that generally refers to active systems.

Solar Energy

Solar power is one of the most common sources of natural energy. In fact, in proper perspective, every energy source is in some way connected to the sun. Distinguishing solar energy from other energy is a bit of a misnomer. Most every atomic particle of matter on this Earth at one time was within a star such as our sun. The sun's energy created the conditions for plant and animal growth. These have led to fossil fuel as a source of energy. Virtually every energy source on Earth is a derivative of solar energy.

Although solar energy could define all energy sources, the one we are most concerned with is the direct gain that will easily convert to electricity. As with some other alternative, natural sources of energy, the availability of direct solar energy is periodic. The sun only shines on any given portion of the Earth about half the time. However, during that limited time, the sun brings us untold treasures of free and abundant energy. Another way the sun converts energy is chemically. We must collect, store, and reuse this chemical energy as best we can. We do so with a chemical process that converts sunlight to electricity called *photovoltaics*.

Photovoltaic Energy

Solar photovoltaic (PV) cells can convert sunlight to electricity through the photoelectric effect that is usable in your home. PV cells, like those in panels illustrated in Figure 11-1, are normally integrated into your electrical system as a supplement. However, many installations

Courtesy of iStock Photo

Figure 11-1 ■ Photovoltaic cells convert solar energy into electricity. One module cannot produce enough electricity for a home, so the modules are linked together to form an array.

are stand-alone systems that produce all the electricity needed for the home. Because the produced electricity is direct current (DC), an inverter is necessary to convert to an alternating current (AC), which is essential for home use. Most commonly there is a need for battery storage to supplement the period when the sun is not available and demand for electricity is normally high.

Conversion of light to electricity occurs at the atomic level. Some materials allow the absorption of light in photons and release electrons. One method of manufacturing the PV cell is with thin wafers of two slightly different types of silicon. One silicon wafer has positively charged electrons. The other type of silicon wafer is negatively charged and has extra electrons. Putting these two dissimilar wafers together establishes a connection that produces a path for electrons when exposed to light. The electron motion follows the path from positive to negative. That motion is an electric current.

Solar PV energy use in your new green home will save energy by generating its own electricity, thus reducing demand on the public electrical system. It will also pay for itself over a relatively short time. Depending on your demand for electricity and the local cost per kilowatt hour, many PV systems have a payback period between 10 and 15 years, based only on the direct savings of dollars not spent on energy. In addition, there are government incentives that reduce the initial cost and shorten the payback period. The Solar Photovoltaic Power Association and the Solar Energy Industries Association are two organizations with an excellent source of information on this form of energy. PV cells can be used in any building. For instance, the apartment buildings depicted in Figure 11-2 offset considerable electrical energy costs with a full roof of PV panels.

Courtesy of iStock Photo

Figure 11-2 ■ Photovoltaic panels are becoming increasingly popular in residential buildings. Even apartment buildings can benefit from free solar energy.

Hydropower

Hydroelectric energy of all types accounts for almost 20% of all energy generated in the world. Small-scale hydropower plants can be powered by the flow of water in a river and supply electricity to your home. Currently over half (60%) of all renewable energy is derived from hydropower through large dams on lakes and rivers. Natural power of this type is not new. Watermills like the one pictured in Figure 11-3 represent mankind's early use of this type of energy.

There are two major types of hydroelectric plants: (1) Conventional, which relies on water flowing in one direction like a river; and (2) pumped storage, which uses water repeatedly, pumping it to a higher storage container and letting gravity bring it down across the waterwheels. Conventional plants that anticipate periodic or seasonal fluctuations in flow rate sometimes store water without the need for a pump. In all cases, the water produces electricity by flowing from a high level to a lower level. The flow rate affects the rate of electrical production. The greater the height (head) and flow rate, the greater the electrical production. There are several ways to make this happen. The earliest version of hydropower use was a waterwheel. It served many purposes, including that of a grain mill as well as a sawmill. Hydropower generators work by diverting a portion of the water and channeling it through a pipe to a generator system that allows the water to be channeled back to the waterway. Electricity is generated by hydropower through a mechanical means. The water flow moves over the turbines or waterwheels and turns the wheel. A generator converts this mechanical energy to DC electricity. An inverter converts this into AC, which is more useful to household appliances and equipment. Transmission power lines, battery backup or storage, and electrical component connections are all part of this type of

Courtesy of iStock Photo

Figure 11-3 ■ Smaller hydroelectrical power generation, although not common, can provide sufficient electricity for most of the demand in a small home. Even micro-hydroelectrical systems, if near a flowing water source, can provide more energy than photovoltaic or wind generation.

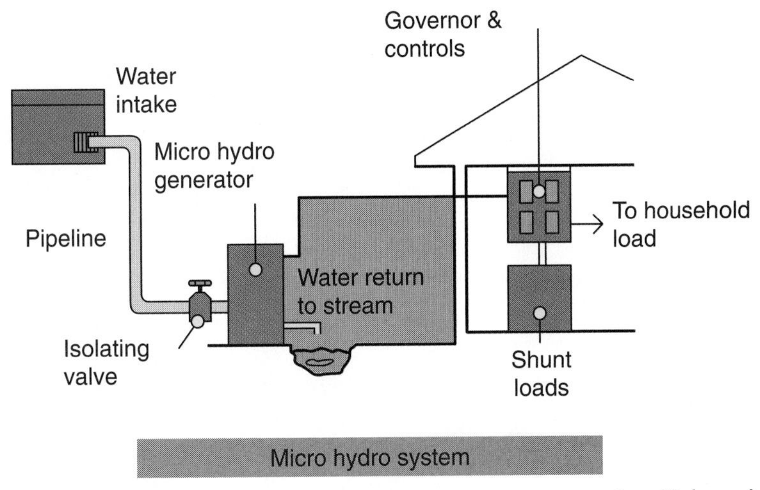

Lynn Underwood

Figure 11-4 ■ Mini- or micro-hydroelectric generators use a flowing water source to provide household electricity while returning the same water to a downstream source, causing virtually no pollution nor disruption to the environment.

assembly. Although complex, it is still appropriate for home use. It is particularly cost effective when located a distance away from an electrical service where the cost is an issue.

However, hydropower applies to more than just large installations. Your green home can have its own system. A micro-hydro generating plant depicted in Figure 11-4 is just right for a single home, provided you are near a reliable flow of water such as a river, stream, or even a creek. Ordinarily, a micro-hydro system uses very little flow rate in a stream. Although there is some variation in the flow rate across the four seasons in a year, normally there is no variation from day to night. Therefore, electrical generation, even on a small scale, will require less battery storage and maintenance for a domestic hydropower plant.

Taken to the extreme, some environmental damage could be the result of excessive use of hydropower. The design of the system is influenced by the flow rate of the pin tube or pipe bringing water to the generator. The flow rate directly affects the electrical production rate. Too much diverted flow, however, can create a concern by adversely affecting the natural conditions of the waterway. If the system diverts too much water into hydroelectric production, the waterway could be diverted and environmental impact could occur. Also, too little water diverted will not provide the needs for the generator.

Wind Energy

Wind energy is a form of solar power in that air over the Earth surface, heated by the sun, causes wind. A wind generator, as shown in Figure 11-5, is a device that collects kinetic energy from the motion of wind and converts it to electrical current mechanically through a generator. Wind velocities and, therefore, wind generator performance increase with tower height. Surface irregularities such as houses, fences, trees, and hills interfere and create turbulence that reduces the generator's effectiveness. Wind speed should be above 10 mph for the generator to work effectively. The usable power created by the generator increases dramatically with an increase in wind speed. Usable wind is available in most areas of North America but is better along coastlines and hills or mountains. Wind patterns and intensity vary along seasonal lines, and wind energy is less predictable than solar energy but available for more hours in the day.

Courtesy of iStock Photo

Figure 11-5 ■ Wind caused by temperature differences near the Earth's surface can be used to generate electricity. Depending on the wind conditions and height of the propellers, most of the electricity needed by a home can be achieved.

Most wind generators operate with a turbine mounted on a horizontal axis. The assembly includes a rotor, frame, generator, and a tail piece that acts as a wind vane. The rotor consists of two or three blades. Having moving parts, the wind generator is a complex machine and subject to maintenance. Depending on the size of the unit, the turbine must be mounted above any obstacle. The larger the turbine, the higher it must be mounted. The wind generator costs money but has a relatively short payback period—normally 10 to 15 years depending on electricity cost and usage. Based on a demand of 10,000 kilowatt-hours per year, a normal size wind generator system for a home would be between 5 and 15 kilowatts. Large-scale wind generators are now in service for residential developments with more on the way. One 600 kilowatt wind turbine can supply hundreds of homes with electricity.

Geothermal Energy

Geothermal energy is derived from beneath the Earth's surface (geo = Earth, therme = heat). Geothermal heat is clean and sustainable. Energy production from the heat below the Earth's surface allows energy to be produced by one of several methods. Direct use of this geothermal heat is possible for heating purposes. Beneath the ground, even as near as 10 feet, the temperature maintains a constant rate near 50°F to 60°F. Geothermal heat pumps can tap into this heat source. These heat pumps have a heat exchanger, a heat pump, and an air handler that work in concert to extract the difference in temperature inside your home and that underground. These work much like a conventional heat pump and the outside air, just changing the source of the heat to underground. In some instances, during the summer months, heat drawn from inside the building can be diverted to supplement water heating needs.

Cogeneration and Micro-Cogeneration

During conventional energy processing, energy is lost through waste. No mechanical device that converts fuel to usable energy does so with 100% efficiency. Some heat is produced and is normally considered to be a loss. This lost energy may be used in a variety of purposes. It can warm buildings or make more electricity. Using the waste heat from power generation in this way is known as cogeneration or combined heat and power (CHP). Cogeneration essentially uses fuel energy two or more times. Combined production of heat and power reduces energy consumption by 20% to 30% compared to separate production. The leftover heat energy that would normally be lost to the atmosphere could be reused as industrial heat, domestic comfort heating, domestic water heating, or even for the generation of electricity. Currently, CHP systems produce almost 8% of the United States' power. This represents billions of dollars in direct energy benefits. It also represents a reduction in energy production of over one trillion (BTUs) every year. Additionally, it reduces emissions and carbon released into the environment. It truly is an environmentally friendly concept.

The concept of cogeneration is more easily explained when considering your automobile. Your car runs using fuel. In the mechanical process of turning a crankshaft, the wheels turn and the car goes forward. But the power generated by the mechanical force is applied to other mechanical and electrical activities. Pressure developed from mechanical systems is applied to other processes through vacuum lines or fan belts. For instance, the air conditioner works by borrowing mechanical power from a fan belt to drive the condenser. The lights, radio, and heater fans all operate from electricity derived from an alternator that is driven by the same mechanical processes fed by a fan belt—the same for the fan that cools the engine. These are examples of cogeneration, using what would otherwise be wasted energy to perform a function.

The use of cogeneration for power is not prolific, but systems are available even for domestic use in your green home. One company, Polar Power, makes a micro-cogeneration unit that boasts the following capabilities:

- 34,000 Btu/hr of air conditioning
- Up to 30,000 Btu/hr for heating hot water
- Up to 6 kW of DC power for battery charging

Or

- In the heating mode it provides up to 36,000 Btu/hr of space heating
- Up to 6 kW of DC power for battery charging
- Up to 30,000 Btu/hr for heating hot water
- The heat from the exhaust can be combined with the heat pump for a total output of 66,000 Btu/hr

A unit like this could heat your home and water and cool your home and supply your electricity as well.

Biomass

Biomass refers to living or formerly living matter that can be processed in some form to create energy through its use as fuel. It can refer to plant or animal matter that is used to produce heat or other energy for industrial use. Wood or straw can be used as fuel for electricity production. Sometimes crops are grown for the specific purpose of biomass fuel. Typical biomass products are wood from trees, planted crops, manure, and some garbage. Burning wood in your fireplace is an example of biomass use. Even certain garbage is used as fuel to heat water, producing steam to create electricity.

Biomass is part of the carbon cycle. Carbon dioxide, which is a by-product of this process, is absorbed by the next generation of trees and plants. Carbon from the atmosphere is converted by photosynthesis into particles of biological material (the plant itself). When the plant is harvested or it decays, the carbon reenters the atmosphere. Because of this, biomass is considered carbon neutral.

Biofuels, such as ethanol and biodiesel, are examples of biomass products. These are generally blended with traditional petroleum products such as gasoline and diesel to reduce the impact of the fossil fuel and the impact on the environment. About 80% of biomass use is attributed to generating electricity. The organic refuse is burned, creating steam that turns generators to make electricity. However, for domestic use, biomass is less than practical, relying on contained combustion of waste matter to avoid contributing to air pollution. There is, however, one use that appears uniquely suited to home heating. Pellets of biomass material are available for use in wood stoves and boilers. Much like paper logs, this biomass product can substitute for traditional wood logs or other forms of organic fuel.

Hydrogen and Fuel Cell Technology

Hydrogen is the simplest element; each atom has only one proton and one electron. It is the most abundant element in the universe. This element has the highest energy content of all common fuels when compared by weight. It is lighter than air and, in nature, exists only as a compound. Hydrogen is found in all living things and, therefore, is part of any biomass.

Hydrogen is used to make or process common products such as ammonia, methanol, gasoline, rocket propellant, and heating oil. It is also used in the manufacturing process for fertilizers, glass, reduction of metallic ores, vitamins, cosmetics, semiconductor circuits, soaps, lubricants, cleaners, and hydrogenation of fats and oils in food products such as margarine and peanut butter. Hydrogen burns readily with oxygen, releasing considerable energy as heat and producing only water as exhaust. It is a clean-burning fuel. Hydrogen can replace today's fuel gas for heating and cooling homes and powering hot water heaters. The problem has been acquiring it in a usable state and storing it safely.

Because hydrogen does not exist naturally as an element, it must be separated for industrial or energy use. There are two basic ways of doing this: steam forming and electrolysis. Steam forming strips hydrogen from methane (CH_4), but doing so contributes unwelcome carbon by-products (CO_2) into the atmosphere as a greenhouse gas. Electrolysis uses electricity to split hydrogen from water into the constituent elements, hydrogen and oxygen. This process results in no emissions but conventional processes are expensive. Photoelectrochemical harvesting is another way of collecting hydrogen. Photoelectrolysis uses PV methods, semiconductors, and an electrolyzer that accumulates and stores hydrogen. Hydrogen can then be used as a fuel.

A fuel cell is an electrochemical energy exchange mechanism. The technology has been known for some time but only implemented in space vehicles in the mid-20th century. Generally available for commercial and industrial applications, fuel cell technology is

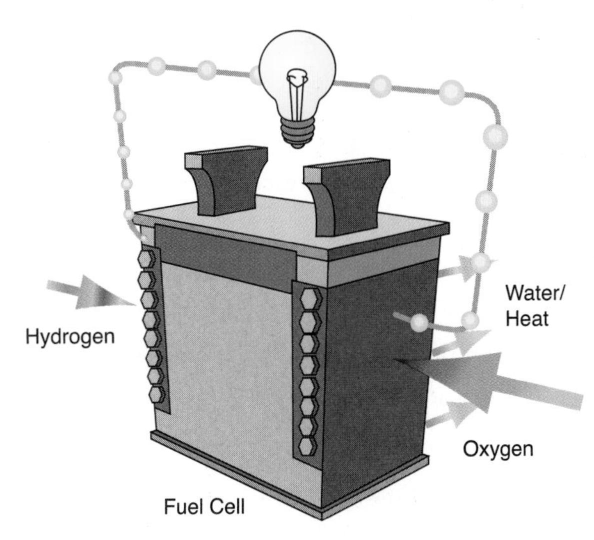

Lynn Underwood

Figure 11-6 ■ A fuel cell.

now available for residential and light commercial applications. Recognized as an emerging technology, provisions are made in the International Codes for its application in home construction.

A fuel cell, as illustrated in Figure 11-6, is much like a battery where electricity is transferred then stored for later use. It uses electricity and chemical reactions to generate and store energy. It can use a chemical such as hydrogen in a pure state or derived from a fossil fuel along with oxygen to generate electricity. In a fuel cell, hydrogen is introduced on one side (the anode side) and an oxidant is introduced on the other (cathode side). These meet each other and chemically react with an electrolyte. The hydrogen is converted into electrons and hydrogen ions. The electrons are repelled by the anode and flow toward the cathode. The cathode attracts the free electrons as well as oxygen, which combine with the hydrogen ions from the anode, producing electricity, heat, and a by-product, pure water.

Several options are available for this energy usage. Hydrogen, as a fuel source, can heat your home and your water. You can cook with it. Wherever you conventionally use fuel oil, hydrogen can take its place. Additionally, hydrogen can help generate electricity in a fuel cell. Many applications for using fuel cell technology are currently being used throughout the world to power buildings, vehicles, and systems using fuel cells.

■ BEFORE YOU DECIDE . . . REFLECTIONS AND CONSIDERATIONS

✔ A green home has a very small environmental footprint

- It does so by conserving energy derived from conventional sources.

✔ Some homes generate their own energy and supplement conventional energy sources.

- Others take the more serious step of generating enough energy to supplement electrical energy or even to be self-sustaining.

✔ There are a variety of ways to achieve the goal of producing your own energy.

- ✔ Although the initial cost of an appliance may be high, it most probably will pay for itself within a reasonable period.
- ✔ Several means of energy generation are available.
  - Solar photovoltaic relies on chemical reactions caused by solar radiation to generate electricity.
    - Good solar access is needed.
    - Batteries are used as a means of storing electricity until needed.
  - Wind generators rely on prevailing wind to generate electricity.
    - Height is needed for wind currents.
    - Batteries are used as a means of storing electricity until needed.
  - Hydropower is widely used as a source of renewable energy but most commonly on a major scale, supplying electricity to entire regions.
  - Geothermal energy devices rely on the stable temperature within the Earth's surface.
  - Micro-cogeneration uses surplus or wasted heat and improves the efficiency of fuel-based equipment.
  - Biomass technology is a broadened effort to take advantage of organic residue that would otherwise rot and decay.
  - Hydrogen and fuel cell technology exploits an energy-rich fuel source that is very abundant and one with harmless by-products.
    - Hydrogen production must be energy conscious and derived from sources that minimize environmental effect.

FOR MORE INFORMATION

Solar Electric Power Association
http://www.solarelectricpower.org/

Solar Energy Industries Association
http://www.seia.org/

Chapter 12

OWNING AND MAINTAINING A GREEN HOME

HOMEOWNER OPERATION

Modern life is complex. The Information Age has allowed us to make huge leaps in technology on a daily basis. We are on a continuous cycle of learning *how to* operate new products and innovations that vastly improve our lives, our society, and our environment. Almost everything we purchase nowadays has complex directions for use. Cell phones, which were unheard of just 20 years ago, now feature hundreds of operations that are accessed or changed by two dozen keypads. Our computers allow us to access or process almost any piece of information we wish.

Your home is no different. Even modern homes that cannot boast the green label are filled with technologies that hail from several aspects of architecture and modern system design, including structural integrity, mechanical equipment, heating and cooling mechanisms, plumbing fixtures, electrical facilities, communications/entertainment equipment, dish and clothing washing facilities, hazard warning devices, emergency egress windows, as well as modern materials that need specific care. All of these appliances, facilities, and innovations require special care as well as regular maintenance. An owner's manual is needed to coordinate the maintenance required throughout your home in an effort to sustain the efficiency and performance you expect.

Manual for the Proper Use and Maintenance of Green Building Systems

Your new green home needs an operator's manual. Usually, the contractor will provide this book to you. If you are building your own home, creating a manual to keep track of important information is crucial. This manual should be filled with all the essential facts about the home and explain the systems built in, their function, and necessary maintenance. The book will probably be between 40 and 60 pages in length and include photographic guidance.

Although the owner's manual will vary depending on particular aspects added to your home, a manual for a green home should cover the following information:

- A complete list of the green building features, their use, and a maintenance schedule.
  - Building, siding, and roofing materials used, interior finish materials used, insulation, windows, doors, weather stripping, air sealing measures, pressurization tests, water-saving devices, plumbing systems, mechanical equipment and water heater efficiency ratings, moisture control techniques, appliance and lighting efficiency, alarm systems (smoke, carbon monoxide, radon), waste management, recycling center, and alternative energy generation.
  - Discussion of air sealing measures and the effectiveness of caulking and the value of proper maintenance of weather stripping. Reference will be made to any repair or maintenance schedule for these installations.
- Product manufacturers' manuals or operating instructions, including any data sheets, MSDS sheets, and warranties for all major equipment and appliances.
- Information on local recycling programs and how to participate.
- Information from a local utility purveyor and any programs that include purchasing a portion of energy from renewable sources.
- Explanation of benefits for using energy-efficient lighting products such as compact fluorescent lamps and light-emitting diodes.
- A detailed list of practices that the homeowner can use for conservation of water and energy.
  - Dripping faucets and running toilets should be repaired immediately and with the most appropriate parts.
- A list of available local or regional public transportation options.
- A diagram indicating the location and operation for connections, safety valves and controls for major building systems such as water, electricity, sewer, storm water drains, HVAC, water heater, range, fireplace, and so on.
- A detailed, photographic record of the building that includes framing techniques, electrical, plumbing, and mechanical installations prior to installing insulation.
- A required maintenance checklist for all aspects of the home, including necessary upkeep of the site such as landscaping.
- A list of hazardous materials used during the construction process as well as those that remain as part of the building.
- Information about organic pest control, fertilizers, de-icers, and cleaning products.
- Information about native landscape materials and Xeriscape that has been planted.
- Information about how to maintain the building's relative humidity between 30% and 60%.
- Instructions for inspecting the building for termite infestation.
- Information about how to maintain gutters and downspouts in good working condition and the importance of diverting rainwater at least 5 feet away from the building's foundation. Information on inspection standards for damage or decay caused by moisture accumulation in various parts of the home.

- Information about proper use of foundation and attic vents.
- Instructions on maintenance of systems that require regular attention such as location of each furnace filter and the estimated replacement cycle.
- The proper setting for the thermostat, its operation, and programming.
- Lighting controls and appliance operation. The most appropriate settings for proper operation for water heater, refrigerator, and freezer. The proper diverter setting for conditioned air outlets.
- Details regarding interior finish materials if low or no volatile organic compounds are used and how to maintain these materials.
- Finally, a copy of the Certificate of Occupancy issued by the building department.

Components of a Green Home

A list of green features should be developed and passed on to the homeowner in a manual. The booklet will cover the value added by each feature and how to maintain its proper working order and functional use. It will have each area outlined previously and developed in complete detail to fully educate the owner. As an example, the following delineates those features for site work and landscaping.

Site and Landscaping

Site and landscaping considerations involve both legal responsibilities and the attributes of a green home. The following serves as a checklist of considerations that affect the site.

- A legal survey is properly filed and a copy of the recorded survey is enclosed. Your ownership is legally established and your deed is properly filed. A copy of the deed is also enclosed.
- You are currently in compliance with all Building Code requirements and zoning ordinance restrictions by the authority having jurisdictions. The following lists some of these rules and regulations:
    - Lot coverage, building setbacks, building size and height, easements, elevation of finished floor, flood plane controls, off-street parking, driveway, exterior architecture, exterior color scheme, and outbuildings may be regulated by zoning. Any changes you want to make require jurisdictional review and approval of your plans before you start.
- Compliance with any land disturbance rule or law.
    - Endangered species, waterway or wetlands protection, rainfall drainage, erosion control, legal access, soil condition, prevailing terrain, landscape and vegetation currently growing may be regulated by the authority having jurisdiction. Any changes require jurisdictional review and approval.
- Xeriscape and landscaping.
    - Certain vegetation has been planted to ensure a pleasing appearance yet is able to survive and flourish with significantly less irrigation than others. Follow the instructions for care of each plant and you will reduce your demand for fresh water. As much as possible, the landscape has been restored to its natural state before construction.

- Sod and lawn care.
    - Watering, cutting, and fertilizing are critical to maintain your lawn's health while still controlling water resources. Follow the directions of the landscape contractor that are enclosed.
- Trees and shrubs.
    - Watering, cutting, and fertilizing are critical to maintain plant health and control wasted water resources. Follow the directions of the landscape contractor that are enclosed. Control weeds and watch for insects and disease.
- Gray water system.
    - You may have a gray water system installed to further conserve water usage. This water is derived from the water you use in your sinks, shower, bathtub, and laundry and is acceptable for use in landscaping only. Maintain this system with the advice of a licensed plumber. Plans for the installation are contained in your house plans.
- Permeable paving.
    - To allow rain to penetrate soil and prevent erosion several areas of paving, including driveway, sidewalks, and stepping stones, are permeable. That is, they allow for drainage. Rainwater will penetrate the paving. If you add nonpermeable paving of concrete work, this will defeat the purpose of the installation. It is recommended that further additions of paving be of the permeable type. A special note: Lot coverage is limited by your zoning ordinance. Adding paving must be approved prior to installation. Note that paving of this type is suitable for medium-sized cars and light trucks.

GREEN HOME RECYCLING CENTER

A new green home simply must have a recycling center to improve the opportunity for the owner to participate in regional recycling efforts. If convenient and easy, the act of recycling will be painless and a part of everyday life. The center should be located where most convenient. A great spot is adjacent to the kitchen where most solid waste comes from. It may be beneficial to have two stages of storage for recycling: short term and long term. Other areas may include the hallway, garage, or a storage area. The short-term recycling collectors can be smaller and, like household garbage cans, be taken to a larger bin when filled. The next step is to select the size of each container. The parameters for this include how the material is collected. If it must be sorted, you will have several containers. If it can be commingled, there can be a larger bin. The size will be limited by what can safely be taken to outside containers or to the pickup location. Crushing cans allows for more recycling space. You can make a crusher out of lumber and a hinge or you can buy one.

An important step for homeowners is to follow the rules established by the recycling center. It may require separation by material: plastic, aluminum, paper, and others. It may require clean items that are rinsed free of food products. It may require glass that is intact and not broken into pieces. It may specify that aluminum cans be compressed. Whatever the conditions, try your best to satisfy them. This ensures that the recycling effort will continue. These specifications are important to include in your manual to help clarify these steps and facilitate recycling practices.

Solid Waste Plan

There are some things that are not accepted at recycling centers yet must be disposed of properly. But not everything must go to the landfill. Consider other options for everything you were going to dispose.

- Hold garage or yard sales for those things you do not want, thus extending material use.
- Give excess material belongings such as clothes to family or friends.
- Give certain effects away to civic or charitable groups.
- Divert yard waste, organic, or (nonmeat) food waste to a compost bin.
- Service stations or mechanics sometimes accept antifreeze, motor oil, or old car batteries.
- Large appliances that are still functioning may be picked up by community organizations such as the Salvation Army. Sometimes large appliance stores will take your old appliance in trade and find the right donation site.
- Dry cleaning stores sometimes accept wire hangers.
- Aluminum, copper, and other metals are sometimes purchased by weight in scrap yards.
- Some music stores will accept your old records and tapes and give you credit for purchases.
- Shelters can use unopened body care products such as soap, shampoo, toothpaste, and so on.
- Some civic organizations accept used eyeglasses for repair and donate them to those who cannot afford eyewear otherwise.
- Hospitals and family centers can use books, magazines, children's toys, games, and stuffed animals.

Household Hazardous Waste Plan

Hazardous materials are present in our everyday lives. The luxuries that we have come to depend on are brought to us through thousands of chemical compounds. Many of the very same chemicals that we fight so hard to prevent from entering our environmentally responsible project are what we bring home from the store and place on the shelf of the kitchen in our green home. It could be anything from pesticides or fertilizers to cologne or toothpaste. We might use it properly and try to keep it from environmental damage but when the container is depleted, we toss it into the garbage that is headed for a landfill while we pat ourselves on the back for being environmentally responsible.

The Resource Conservation and Recovery Act (RCRA) is a federal law that was passed in 1976 to protect the public from harm due to waste disposal. The broad category for disposal of waste that is harmful but disposed of by residences is called Household Hazardous Waste (HHW). Because regulating every garbage can in the United States is an impossible task, homeowners sometimes dispose of many types of household hazardous wastes by putting them in their trash. Besides putting it in the garbage, improper disposal of household hazardous wastes is done by pouring them down the drain into the sewer system, pouring them on the ground, and pouring them into storm sewers. There are plenty of solutions to the unsafe disposal of these household hazardous wastes. First, if you can find an alternative material that is

Table 12-1

An Abbreviated List of Common Household Wastes That Could Be Considered Hazardous

Cleaning Products
- Oven cleaners
- Drain cleaners
- Wood and metal cleaners and polishes
- Toilet cleaners
- Tub, tile, shower cleaners
- Bleach (laundry)
- Pool chemicals

Automotive Products
- Motor oil
- Fuel additives
- Carburetor and fuel injection cleaners
- Air-conditioning refrigerants
- Starter fluids
- Automotive batteries
- Transmission and brake fluid
- Antifreeze

Lawn and Garden Products
- Herbicides
- Insecticides
- Fungicides/wood preservatives

Other Flammable Products
- Propane tanks and other compressed gas cylinders
- Kerosene
- Home heating oil
- Diesel fuel
- Gas/oil mix
- Lighter fluid

Indoor Pesticides
- Ant sprays and baits
- Cockroach sprays and baits
- Flea repellents and shampoos
- Bug sprays
- Houseplant insecticides
- Moth repellents
- Mouse and rat poisons and baits

Workshop/Painting Supplies
- Adhesives and glues
- Furniture strippers
- Oil or enamel based paint
- Stains and finishes
- Paint thinners and turpentine
- Paint strippers and removers
- Photographic chemicals
- Fixatives and other solvents

Miscellaneous
- Batteries
- Mercury thermostats or thermometers
- Fluorescent light bulbs
- Driveway sealer

Source: Environmental Protection Agency

not regarded as hazardous, buy it instead, and decrease the demand for the hazardous materials. This will eventually result in the decline of production and distribution of hazardous products. Avoiding certain items will be easy with some products and impossible with other products. According to the Environmental Protection Agency, the items in Table 12-1 represent the most common household hazardous wastes.

A danger exists if these materials are introduced into our community solid waste disposal sites. Although federal law does not criminalize this type of disposal, many state, regional, or city solid waste facilities offer special pickups for proper disposal that brings less harm to the environment. The best advice is to call your local environmental, health, or solid waste agency for information on proper disposal of these household hazardous waste materials.

ENERGY USAGE FEEDBACK

A green home relies on information feedback on system usage to alert you to arising conditions that affect the performance of your home. The more efficient the operations of the systems, the more productive your green home will be at saving energy and avoiding waste. In part, a manual on homeowner operations serves that purpose, reminding you of the myriad of components that need service or repair (or even replacement) over time. Modern technology has now taken this a step further with an energy monitoring dashboard that gives you real-time feedback on your energy usage. Imagine watching the energy consumption in your home from an electronic component mounted on the wall much like a thermostat. That technology is here right now. General Electric has produced a sophisticated device that monitors and reports energy usage.

The wall-mounted monitoring system not only allows you to watch current and historical energy usage, but it also serves as a programmable thermostat, audio control for MP3 players, security system, intercom, lighting control, and monitor for solar power usage. It is a significant part of the new green home because it gives you the full picture of energy usage so you can make necessary changes.

SAMPLE HOMEOWNER OPERATION MANUAL

Building Green in New Mexico has an example of a homeowner operation manual that could serve as a guide for you. This homeowner's manual includes monitoring cycles, operation, maintenance, and repair schedules for various appliances, equipment, and systems as well as reminder tips for water usage, recycling, and disposal of hazardous waste. It includes indoor air quality and energy use optimization considerations, tips on alternative transportation, and utility shutoff suggestions. http://www.buildgreennm.com/downloads/homeownersmanual.doc/

BEFORE YOU DECIDE . . . REFLECTIONS AND CONSIDERATIONS

- ✔ Modern homes are complex structures filled with even more complex systems that need maintenance to run properly.
    - A green home is even more complex in every way. A user's guidebook or operations instruction manual is essential to allow the owner/operator to fully understand every feature present in the new green home.
- ✔ A homeowner's manual should have how-to guides for facility systems and mechanical operations.
    - It will include a detailed maintenance schedule and key things to note and correct if necessary.
    - It will have a complete list of all the installed components that make the house environmentally friendly. It will further explain how each particular component is friendly to nature.

    - The guide will delineate practices that lend toward conservation and energy savings in all of the building's components.
- ✔ In addition, a green home must have a recycling center that makes it convenient to save certain materials and participate in regional or city-wide recycling programs.
    - Also, an established pattern for recycling and proper disposal of solid waste must be established and practiced as a routine for the owners of the green home.
- ✔ An energy usage dashboard allows you real-time monitoring of current and historical energy usage so you can adjust or tune up systems and equipment, or make lifestyle changes that reduce energy usage.

FOR MORE INFORMATION

Building Green in New Mexico—a sample homeowner's manual
http://www.buildgreennm.com/downloads/homeownersmanual.doc/

General Electric—energy monitoring dashboard
http://www.ge.com/yourhome/dashboard.html/

INDEX

('f' indicates a figure; 't' indicates a table)

A

B

C

D

E

F

G

M

N

O

P–Q

R

S

T

U

V

W–X